Treasure
Caches
Can Be Found

Charles Garrett

Above: The dream of every cache hunter, to find a century-old stagecoach safe complete with its long lost gold, silver and numerous personal items. This cache was recovered from a cellar located in the middle of a field. The treasure hunter's metal detector sounded off loud and clear indicating a very large metal item had been discovered. Upon digging, the finder unearthed several pieces of roofing corrugation beneath which was the safe, where it had obviously been hidden many decades prior to its discovery.

COVER PHOTO: This earthen jar of thousand year old coins was discovered on an ancient European battlefield and is Charles Garrett's favorite cache story. When Garrett first learned of its discovery, he knew he must display and tell the story of this "treasure of treasures". It is believed these coins were hidden by a military commander shortly before a life or death conflict. Clearly, the cache was not recovered by its owner and stands as a testament that there are still thousands, even tens of thousands, of caches still waiting to be found.

ISBN 0-915920-93-X
Library of Congress Catalog Card No. 2001048914
Treasure Caches Can be Found
©Copyright 2004
Charles L. Garrett

First Printing, September 2004

Dewey Reference Number: G525 .G6923 2002
                            622'.19'0973--dc21

Library of Congress Cataloging-in-Publication Data:
Garrett, Charles L.
    Treasure Caches Can Be Found:
    by Charles L. Garrett
        p. cm.
    ISBN-0-915920-93-X
        1. Treasure-trove.
        2. Treasure-trove-United States. 1. Title

For a FREE listing of related treasure hunting books write: RAM Publishing Company, 1881 W. State Street, Garland, Texas 75042 or visit us at www.garrett.com.

Charles Garrett and other professional treasure hunters give you an in-depth analysis, interspersed with exciting hints, useful tips and bold insights about the rewarding adventures of treasure hunting in these fine books. Each book holds a treasure of fine color and black & white photos as well as important treasure hunting references.

Ghost Town Treasures: Ruins, Relics and Riches
Clear explanations on searching ghost towns and deserted structures for caches and many other treasures; which detectors to use and how to use them. By Charles Garrett

Find An Ounce of Gold a Day
Pocket Guide provides basic instructions on using Gravity Trap pans to recover gold. By Roy Lagal

Introduction to Metal Detectors
Pocket Guide carefully explains, in layman's terms, how metal detectors operate and how to find treasure with them. By Charles Garrett

Treasure Hunting for Fun and Profit
Basic introduction to treasure hunting with a computerized 21st-century metal detector; a "must" for beginners, interesting reading for old-timers. By Charles Garrett

Modern Metal Detectors
Comprehensive guide to metal detectors; designed to increase understanding and expertise about all aspects of these electronic marvels. By Charles Garrett

Treasure Recovery from Sand and Sea
Step-by-step instructions for locating and reaching the "blanket of wealth" beneath nearby sands and under the world's waters; rewritten for the 21st Century. By Charles Garrett

You Can Find Gold. . .With a Metal Detector
Explains in layman's terms how to use a modern detector to find gold nuggets and veins; includes instructions for panning and dredging. By Charles Garrett and Roy Lagal

Gold of the Americas
A history of gold and how the precious metal shaped the history of the Americas; filled with colorful vignettes and stories of bravery and greed. By Jennifer Marx

The New Gold Panning is Easy
Excellent field guide shows the beginner exactly how to find and pan gold; follow these instructions and perform as well as any professional. By Roy Lagal

Buried Treasures You Can Find
Complete field guide for finding treasure; includes state-by-state listing of thousands of sites where treasure is believed to exist. By Robert Marx

The Competitive Treasure Hunt
How to plan, organize and win competition treasure hunts. By Jack Lowry

The New Successful Coin Hunting
The world's most authoritative guide to finding valuable coins, includes information on computerized detectors. By Charles Garrett

Sunken Treasure: How to Find It
One of the world's foremost underwater salvors shares a lifetime's experience in locating and recovering treasure from deep beneath the sea. By Robert Marx

Treasure Caches: Can Be Found
Charles Garrett explores the fascinating hobby of cache hunting and teaches you how to properly research, locate and recover treasure caches from decades and centuries past. By Charles Garrett.

To order a RAM book call 1-800-527-4011 or visit our web site www.garrett.com for more information.

Charles Garrett and other professional treasure hunters bring the hobby of metal detecting to life in these fun-filled treasure hunting videos. From the canyons of Mexico to the waters of the Indian Ocean, you will experience the thrill of treasure hunting with the world's most experienced treasure hunters.

Southwestern Treasures
Treasure of Mexico - Electronic Prospecting by the Garrett field team in Mexico's Cobre Canyon; Gold and Treasure Adventures - Competition hunt in California; Treasure hunting in Europe.

Treasure Adventures
The Silent Past - The ghost of a Big Bend prospector helps a modern-day family discover buried artifacts; Tracking Outlaw Treasure - A modern day search of cached loot from a 19th Century stagecoach robbery.

Weekend Prospecting
Shows all current techniques for locating gold nuggets, veins, specimens and placer deposits with metal detectors and gold pans.

Sand and Sea Treasures
Treasure Recovery from Sand and Sea - Slide show on video. "How-to" beach/surf/water hunting tips; Treasure of Indian Ocean - Search for 18th Century wreck off Madagascar shows underwater recovery.

Gold Panning is Easy
Tells where and how to locate gold with the world-famous Gravity Trap gold pans. Step-by-step gold recovery techniques from companion RAM books are shown.

Treasure Visions and a Utah Treasure Trek
Charles Garrett documents ghost town and other hermit cabin searches and actual finds of several caches of gold, silver and coins. Techniques from companion Ghost Town Treasures (book) are show in the award-winning video. A Utah Treasure Trek is a 20 minute behind-the-scenes filming of Treasure Visions that runs immediately after Treasure Visions.

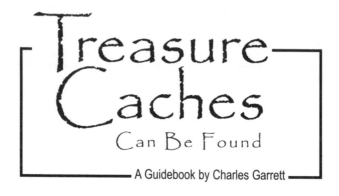

Treasure Caches
Can Be Found

A Guidebook by Charles Garrett

Dedicated to those
who believe that
hide-n-seek is not
a mere child's game,
but a lifetime pursuit
of adventure, discovery,
intrigue and of course,
abandoned treasure.

Table of Contents

About the Author

When only six-years-old, Charles Garrett buried a cache of shiny buttons beneath his home in the pinewood forests of East Texas. Today, that button cache remains where young Garrett buried it so many years ago. No doubt it will one day make an exciting find for a modern treasure hunter, who very well may be using a Garrett metal detector!

Over the past 60 years, Garrett's passion has focused on finding caches from decades, centuries and millennia past. This book was written, among many reasons, to share his prolific documentation and personal search of hidden treasures with hobbyists around the world.

Of course, Garrett did not set out to become the world's foremost expert of treasure recovery as well as law enforcement and security metal detection, nor to become a worldwide manufacturer of metal detection products. However, his passion and enthusiasm for metal detection has earned him this great distinction. His life is continually dedicated to the pursuit of new techniques and products that improve the possibilities of finding America's hidden treasures and improving security screening checkpoints.

Following high school, Garrett enlisted in the U.S. Navy where he advanced to the rank of Electrician Mate 2nd Class Petty Officer. It was during a six-month Electrician's Mate School that his education in electronics began. After his service in the Navy, Garrett earned his Bachelors of Science degree in electrical engineering at Lamar University in Beaumont, Texas. He then began working for Texas Instruments in Dallas in 1959. There he played an integral part of the company's Space and Electronics division where he helped design solid state amplifiers and power supplies for a data encoder which was used in the Mars-bound Mariner II spacecraft. Later, he joined Teledyne Geotech, in Garland, Texas. There he became heavily involved in several earth science related projects, as well as the advanced seismograph circuitry that was planted on the moon by Neil Armstrong to detect moon quakes and other seismic disturbances.

During this time, Garrett was also pursuing his passion of metal detection. Each night after work, he sat in his small, garage workshop where he worked tirelessly on building a better metal detector. By 1964, he had developed The Hunter, which revolutionized the market with its Zero Drift and dual searchcoil technology. Today, Garrett Metal Detectors' products are heralded around the world for serving as the catalyst for locating amazing treasures as well as becoming a pivotal influence in the development of metal detection screening checkpoints in security applications worldwide.

After more than a half century, Garrett has earned countless awards and recognitions from international, national, state and local governments, dignitaries, clubs and individuals who have praised his contributions to the development of metal detection technology. An accomplished author and renowned spokesperson, Garrett continues to actively serve in his role as President and CEO of Garrett Metal Detectors.

He is married to Eleanor Smith of Pennington, Texas, who plays a major role in the development, growth and success of Garrett Metal Detectors. They have two sons and a daughter.

Over the past 40 years, the Garretts have set the standards of design and application for metal detection worldwide. And they are responsible for bringing the treasure hunting hobby to hundreds of thousands around the world.

From the beginning, Garrett has "practiced what he preached'" by personally field testing his equipment to ensure it performs under the most demanding conditions. He has spanned six continents as well as many of the world's seas and lakes in search of lost or hidden treasure.

This book is his personal invitation to join him "in the field" as he searches for caches and recounts the tales of some of America's most infamous outlaws and "shady" characters whose hidden loot may still be waiting discovery.

You are also invited to visit The Garrett Museum of Metal Detection Treasures located in the company's corporate offices in Garland, Texas. There you will be able to witness, firsthand, the excitement that comes from the fascinating hobby of metal detection.

Aaron Cook
Hal Dawson
RAM Editors

Author's Note

The study of history is a colorful and exciting adventure that has the power to excite and inspire our imagination. And that's why I wrote this book. America's past is riddled with colorful tales of notorious outlaws and shady characters whose legends inspire and motivate us to delve more deeply into our history. In addition to looking into the past through the eyes of a treasure hunter, this book provides precise, step-by-step instructions for researching, seeking and recovering caches of all kinds with a metal detector. From the Rocky Mountains to the sandy sea shores, there are countless lost treasures that still lie in wait, undisturbed, for the person who is willing to research, study and work hard to recover them.

Discovering hidden treasures is the dream of every metal detector hobbyist, whether he or she will admit it or not. What could be more exciting and rewarding than researching, hunting and locating money, jewels and other valuables hidden eons ago by some desperado?

Hunting for treasure with a metal detector is a hobby filled with romance and adventure. Does discovering a button from the uniform of a Civil War soldier, the badge or gun of a 19th Century lawman, a gold nugget in a mining camp or an Indian head penny stir your imagination?

Many agree that cache hunting is the most romantic and potentially the most profitable form of metal detecting. When looking for a cache, you're searching into the past, which is truly romantic and exciting. If your cache is loot abandoned by an outlaw, the hunt becomes even more compelling and may be far more rewarding.

Believe me, with a modern metal detector, you *can* find a cache. It might not be the big one hidden by Frank and Jesse, Butch and Sundance, John Dillinger or other notorious outlaws described in this book, but you can find one. Even if it's smaller, searching for it is no less romantic.

Okay, you say. Tell me about it. What kind of cache are you so certain I can find?

It could be a fruit jar filled with coins long ago buried in an abandoned garden; a purse with a few bills hidden in the closet of an abandoned house; even a tobacco tin full of bills hidden within the wall of a dilapidated, old dugout. These are typical caches that still lie in wait for you to recover.

The point is, there are caches of all sizes out there, just waiting to be found. I know because I, along with many of my treasure hunting colleagues, have found several of them. Many are discussed principally in Chapters One through Seven of this book. These chapters will spawn ideas of just what caches are, where they're located and specific instructions on recovering them. Often I give you treasure hunting clues and if you're paying attention, you'll learn a little more about my treasure hunting techniques.

In the final chapters we'll explore the real romance and adventure of cache hunting and review some legends that feature famous bandits and their caches. I'll admit that some of the caches attributed to these renegades may not exist. A great deal of folk lore has grown around the exploits of Sam Bass, The Wild Bunch, Jean Lafitte and others whose stories I present.

Thus, I urge you to carefully follow the instructions outlined in Chapter Three before spending much time, money and other personal resources searching for outlaw caches. Reading about these exciting treasures and how they came to be hidden by bandits long ago takes little time and no money. It certainly sets my mind to wondering into the past as well as my prospects for digging one of the great hidden outlaw treasures.

But, make no mistake about it, many outlaw caches existed at one time. It's a well known fact that banks, stagecoaches and trains were regular targets by Old West bandits and that much of their loot was never located when, or if, they were apprehended during their next robbery. It only makes sense they hid it somewhere!

Maybe a gang member recovered it the next day. Maybe it was found accidentally a few months or years ago. Or, maybe a treasure hunter, just like you, found it last week. Who knows, the cache maybe still be out there waiting for us.

Will this book tell you exactly where the outlaw treasures are hidden? Of course not. If I knew exactly where they were, I'd have dug them up myself long ago. But, I've certainly spent a great deal of my life in the field looking for, and finding, treasures.

You may even be wondering if I've ever discovered an outlaw cache myself. The answer is yes, perhaps several. However, my "yes" depends on your definition of "outlaw". That may seem to be an evasive answer, but later in this book I'll tell you of some of my exciting adventures. I'll also let you in on reliable information on many other caches. In fact, I'm on the trail of several right now. Let me quickly admit, however, that over the years I've found a lot more gold, silver and jewelry than outlaw caches - most of which I found while searching for caches.

Since this book contains a compact summary of the knowledge I've accumulated about caches during a lifetime of treasure hunting, you may note that some of it has been presented in my other writings. Yet, this book offers an expanded treasure trove of cache hunting information. I continually add to my knowledge through personal experience and contact with other veteran cache hunters. In this book you'll learn of my successes as well as mistakes. Both are vital to successful cache recovery.

If you're a dedicated treasure hunter like me, I know that you'll enjoy reading these instructions about finding caches. And I hope you set on the path to a bandit's treasure, if seeking one is truly your aim. Remember, however, that while it's certainly an admirable goal, I strongly recommended that you set your sights on smaller caches. You'll be amazed at how many there still are. This book offers solid recommendations about where and how to find them. In fact, the owners of thousands of hidden or lost caches are still alive and may even want to help you find their lost treasure.

After six decades of treasure hunting, I've learned one truth that I readily share with you: The treasure is out there. It's up to us to find it.

It has taken me ten years to assemble the material and write this book. I truly loved doing it as much as I loved every minute during each one of my cache searches. I wish I could have bought this book 60 years ago and put it to good use. I assure you, as a cache hunter, I would be saying, "the cost of *Treasure Caches Can Be Found*, was the best investment I ever made." I also suggest you read, *Buried Treasures You Can Find*, written by Robert Marx and published by RAM Books. Look for the RAM Books order form included at the back of this book.

Chapter One
What is Cache Hunting?

If you haven't experienced the enthusiasm of hunting and finding caches with a metal detector, you don't know what you've been missing. True, other forms of the treasure hunting hobby give you the exciting joy of discovery and the benefits of relaxation, fresh air and outdoor exercise. But, you'll find the search for a cache, the thrill of the hunt, can be just as rewarding as finding the cache itself.

I certainly don't want you to believe there's a particular "magic" to cache hunting. It's just another aspect of the wonderful hobby of seeking treasure with a metal detector.

But, always remember, no matter how successful you've been at finding coins or jewelry, cache hunting is different.

You may ask, aren't all treasure hunting books designed to help me find coins in the park, jewelry at the beach and relics at a battlefield? Yes, finding wealth of some type should be one of your goals whenever you go treasure hunting, along with other equally valuable goals.

But this book explains how you must think and act differently when you hunt for caches. Always remember, you'll be looking for big (relatively speaking) money that someone intentionally hid from you as well as trying to think like that person at the time he or she hid their wealth. True, it will help to use all knowledge you've developed in other kinds of hunting, both with and without a metal detector. However, it's your overall manner of searching, from research to recovery, that must be different if you are to be a successful cache hunter.

Successful cache hunters are a dedicated breed and their focus pays off in tangible rewards. They are willing to overcome obstacles because they seek real treasure, financial wealth, and a bundle of it. In fact, dedicated cache hunters gladly welcome obstacles that limit the number of hobbyists searching for the same prize.

Of course, even good cache hunters aren't successful every time. The beginner should realize this and not become discouraged. Always remember, there are literally millions of dollars stashed in the ground (or ocean) waiting to be found. If you persist, sooner or later you can hit a cache. It may be only a few dollars tucked into a tobacco can. Then again, some treasure hunters have become wealthy from pursuing this fascinating hobby and lots of treasure has been found by accident; perhaps more than found with a metal detector.

Looking for money caches generally means searching for a large quantity of buried treasure. Your cache can be an iron kettle filled with gold or silver coins. It can be a cache of gold or silver bars or even guns. You'll generally be looking for objects much larger than single coins, relics or nuggets, though small money caches in coin purses and tobacco cans are often found.

As I have pointed out, many cache stories sound too good to be true because the facts are often stretched and the stories are hard to prove. I know, however, that the following story is real.

My colleagues searched a site in Europe near the shores of the Adriatic Sea and were delighted to find countless individual coins and relics. Upon further investigation, however, one cache hunter's detector sang out loud and clear. It was obvious that he discovered a very large metal object.

It turned out that he found what I believe to be one of the largest
18

treasures ever discovered with a metal detector in Europe. In fact, it required three trips for the entire horde to be transported in a single automobile. Just the weight alone was almost unmanageable.

What did he find? Ah, that's the romance of cache hunting. I'm not going to tell and I don't believe anyone else will either. I will say that the discovery enriched the individuals who made it in more ways than one.

Because cache hunting is different, the basic concepts governing it are also different than those of other forms of treasure hunting. Yes, you should remember and follow the basic treasure hunting

ABOVE: Often treasure hunters send photographs of coins, relics and caches they have found to the Garrett factory. We thank the lucky person who sent this cache photo. It encourages all of us to continue, unwavingly, in our quest for lost treasures.

rules that you have learned over the years. By all means, use those special hunting methods that have proved successful for you and your detector. As I continue to emphasize in all of my books and articles, basic techniques of metal detecting remain the same because the laws of physics do not change. Whether you can be successful in cache hunting, however, will to a large extent be determined by the manner in which you apply basic techniques and how you utilize the technology of modern detectors. Even though I urge you to trust your experience, I also recommend that you consider these guidelines that have proven successful for many cache hunters.

- **Research** - Conduct extensive research. You can never know too much about your target, the individuals who hid it and the circumstances surrounding it.

- **Be Patient** - Develop patience throughout your search. From planning to scanning and recovery, even after you dig up your prize, patience will add immeasurable value to your hunt.

- **Don't Assume** - Never assume that just because your target is supposed to be big, that it will be easy to find. Sure, some caches are quite large. However, they may be deep and consequently more difficult to detect.

- **Don't Anticipate** - Don't presume anything about the cache itself or the area where it is supposed to be located. You can be certain its resting place is not as you imagined.

Now, let's consider additional, important factors of basic treasure hunting techniques. Each of these considerations will enter into the successful recovery of a deeply buried cache:

20

- **Geographic Location** - Know the specific treasure site. Don't rely on a good guess. You must located the site before you start scanning for the cache.

- **Ground Condition** - Be prepared for the terrain and vegetation (or lack thereof) of the treasure site. Be prepared to cut weeds.

- **Physical Changes** - Nature and man change constantly. Keep in mind that physical changes in the environments have probably occurred at the site since the cache was buried. Fill dirt may have piled up, or a trash dump may now be there. Natural changes may be even more dramatic. Layers of silt may cover it more deeply or erosion may have changed the appearance of the location itself.

- **Mineral Content** - The soil's mineral content can have a significant impact on a metal detector. Know how to use your metal detector to overcome mineralized soils.

- **Cache Size** - Unfortunately, the actual size of the cache is often exaggerated. So, remember that any size searchcoil can find both small and large caches if they are buried at shallow depths. And remember, there may be multiple caches at one site.

- **Cache Depth** - It's best to use the largest searchcoil available for your detector or a deep-seeking two-box coil. In fact, I strongly urge you to use nothing smaller than a 12" searchcoil in most of your cache searches.

- **Detector and Searchcoil** - Tremendous improvements have been made in metal detection technology over recent years. Take time to understand what they are

21

ABOVE: Charles Garrett and his family spend a great portion of their time enjoying beautiful mountainous country. Charles takes advantage of these wonderful times by searching for gold nuggets, caches and relics. And hardly is there a mountainous visit that he does not test new Garrett equipment to further prove its capabilities to operate flawlessly over rough terrain that contains ground mineralization. Notice that he is using his favorite hard case backpack. These backpacks will easily hold a complete metal detector and provide perfect protection even if Charles might fall down a mountain, which he has - several times.

and take full advantage of them.

Misjudgment of any one of these guidelines can keep you from recovering the prize(s) you seek. Experienced cache hunters always make allowances for the condition of the search area and the fact that their cache may be smaller or buried deeper than anticipated. Pay close attention to the description(s) of exactly where you believe (based on research) it was hidden. And, when you reach the probable location of your cache, don't rule out a site because of its present appearance. So what if it doesn't look like the description written decades ago or centuries ago. Buildings and other structures and even rocks and boulders may no longer exist. Men could have torn down and hauled them away. Remember that trees and shrubs grow taller, die or are removed entirely. Plus, never underestimated the effects of erosion and sediment. What was once a deep ditch may be just a depression today and vice-versa.

Take time. Be patient. Reap the rewards.

In later chapters we'll discuss research and recovery techniques as well as recommended equipment for cache hunting. First, however, let's look for a moment at the true cache hunter, that "different" breed in our treasure hunting fraternity.

The cache hunter is seldom seen among weekend hobbyists, those who who hunt coins and relics just for the fun of it. The cache hunter spends time in pursuit of large, more profitable finds.

An interesting cache story, illustrating such a search was told in my novel about the infamous John Murrell, whom you'll learn more about later in this book. Based on information given me and personal investigation, it's evident that John Murrell had cached stolen loot. I organized a group to pursue this treasure story and

I believe we located its underground location. When we finally began to dig, however, all that was left was a fresh hole in the ground. One of our group had preceded us and dug up what we had researched and located. I have no idea what he found, but I believe I know who did it.

For a fun, fictional story about our group's adventure searching for Murrell's cache, you can read *The Secret of John Murrell's Vault* printed by RAM Publishing Co.

So, we didn't find the treasure. But, was the hunt a complete waste? Of course not. The motives of a cache hunter are never entirely financial. Nor, does the cache hunter consider hunting for coins or small relics to be beneath him. Cache hunters seek "something big" just as the big game fisherman is willing to spend a lifetime in pursuit of that record-breaking marlin or sailfish.

More than a decade ago I described in a book, what I believed to be the basic characteristics of a successful treasure hunter. Several of them are particularly applicable to cache hunters.

- **Desire** - While the desire to find lost treasure, the belief that lost treasure is waiting to be found and the curiosity / greed to find it, is basic to any treasure hunter, it is particularly apparent in good cache hunters.

- **Love of Historical Knowledge** - Successful cache hunters are thrilled to delve into research that might uncover a treasure lost for years. They enjoy reading historical books and histories of events at all levels - local as well as national. They seek information about the past, attempting to take a single fact and begin the journey that ends with buried treasure. Sometimes it's as obvious as "a wealthy man farmed here, yet he never had a bank account and died without leaving anything

24

but the farm itself." More often, it's simply, "the loot was never recovered." From this single fact, or rumor, they seek additional information until a full scenario is developed and they need only follow the trail with their metal detector to discover where to dig.

- **Patience** - The simple virtue of patience presents obvious rewards that is required by all successful treasure hunters who, believe me, come to know that many disappointments generally precede striking it rich. The successful treasure hunter works to transform the fruitless efforts of today into building blocks or stepping stones that will lead to tomorrow's caches.

- **Analytical** - Successful cache hunters enjoy getting caught up in the details of a cache story. This makes them more prepared for the hunt and brings them closer to the cache before they even begin to dig.

- **Travel** - An accomplished cache hunter must always be ready for travel. There's no telling where a good treasure trip might lead you and your trusty metal detector.

- **Hard Work** - Like any hobbyist who wants to be successful, a good cache hunter must have no fear of hard, physical labor.

- **Equipment Savvy** - By this I mean, that successful cache hunters are especially interested in metal detection equipment. It may include any type of detector or other device that will make it easier to locate that prize that somehow eluded others. Now, this doesn't mean that an outstanding detector or any other piece of equipment will cover up flaws in a cache

25

hunter's character. But, the most effective cache hunter absolutely requires the best equipment possible. You can be certain of that.

- **Human Nature and Greed** - I list this one last, not because it is least important, but because it may very well become your most important rule. By placing it last, this rule will hopefully remain the freshest in your mind.

  Greed drives people to steal things they are not willing to work for, even if it means stealing from a friend, employer or even a stranger. Just give them the opportunity to steal something they covet and they will take it. Human nature works in strange ways and you should remember this. As soon as a person steals, he or she will exchange it for something else they wanted or will try to hide it from anyone who might try to steal if from them. The best, personal example I can offer about how another person's human nature and greed can work for you is explained in the following true story:

In the early 1980s, some colleagues and I searched for valuable, discarded pieces of silver and had been quite successful. It came time for lunch however, so my treasure hunting companion, Roy Lagal, and I decided to rest and eat in the shade of the large towers that the early-day miners used to bring their silver up from the mines.

As we ate, we talked about various treasure hunting methods and the conversation turned to high grading gold and silver from the mines. It was a common practice of early miners, who picked their way through the veins in search of precious ore, to steal a piece now and then, which they hid in their clothes or even mouth. And when the opportunity presented itself, they hid it in a location

26

TOP and BOTTOM: These photos were taken during the actual recovery of two silver ore caches that were put down possibly 70 years before Garrett found them. Garrett's instincts of human nature led him to search and find high graded silver ore near this large rock. Garrett refers to this cache as his "Double Las Vegas Jackpot", which you can read more about on these pages.

Silver Cache   Silver Cache

3 feet tall
18 inches wide
egg shaped rock

ABOVE Illustration depicting how Garrett's Double Las Vegas Jackpot was cached by an early day miner. BELOW: Illustration depicting the hole left when Garrett removed the rock and how the high graded silver fell into the hole resulting in the sound of a Las Vegas jackpot machine.

ABOVE and BELOW: Once flourishing silver mines, which are now abandoned, were once home to the "Double Las Vegas Jackpot" cache of high graded silver ore found by Charles Garrett in the early 1980s.

for later recovery. This subject had been on my mind for quite some time and we even discussed the possibility that there might be some of these caches nearby, waiting for someone with a metal detector.

As I sat there and visually scanned the area, I saw a large rock protruding from the ground that resembled the top of a Volkswagen car and noted to Roy that there might be silver buried near it. He responded simply, "why don't you go and see." That was all the encouragement I needed to get back to metal detecting. I walked over to the large rock and immediately began scanning around its base, searching for a signal. Suddenly my detector made a very peculiar sound. At first, I thought the sound was mineralization that was giving me a false reading. I scanned several more times trying to pinpoint the exact location and finally decided the best way to make a decision would be to begin digging.

The large rock was mineralized, so it was difficult to exactly pinpoint the target. But, I was about to find out what made pinpointing so difficult. I dug a wide, shallow hole and scanned again. I received several very faint signals, but nothing told me to "dig here!" So, as I began digging, removing the top soil, my shovel suddenly struck the top of a rock. As I removed the soil to about 8 inches, the rock began to take on the shape of a round, five gallon gas can. My detector told me there were two metal objects, one on either side of the 'gas can' rock.

Because of the dual signals, I immediately perceived that someone buried a very large piece of metal, possibly even a safe or money cache, and placed the 'gas can' rock on top to discourage anyone who might dig here. And too, my detector's erratic signals could possible have been caused by the mineralized rock that was sitting square on top of this 'treasure.' I bet you can see why I was confused.

ABOVE: Garrett examines a common cache container, the hermit Prince Albert tobacco can. A variety of caches, including gold and silver coins, nickels, pennies and marbles were hidden in these types of cans. Hundreds, perhaps thousands, of these types of caches still exist because they were commonly used to hide an individual's treasure.

ABOVE: Hobbyist in the United Kingdom compares coins recovered in a cache against published descriptions to determine their precise value.

I decided to dig the rock out. I dug the soil in front of the rock, grabbed it's top, pulled outward and rolled it down the slope, exposing a deep hole where the rock had been. I felt down in the hole, but found nothing but dirt. I grabbed my detector, inserted the searchcoil into the hole and scanned again. Immediately the detector rang out with a very large signal; the kind I like to say, 'nearly blew out my detector's speaker cone!' I rotated the searchcoil around, but only heard a continuous loud sound.

I adjusted my audio threshold until it was very faint then rotated the searchcoil around the hole's sides. Suddenly I heard a very loud signal telling me, 'this is it, dig!' I grabbed my hunting knife (which I take on all of my treasure hunting trips and nicknamed "Buddy") and jabbed the blade into the side of the hole exactly where my detector said, "dig."

Suddenly, I heard a series of metallic sounds that sounded exactly like a Las Vegas slot machine paying off in silver dollars! I truly could not believe my ears as I kept hearing that metallic "tinkeling" sound coming from the hole. When the "slot machine" finally stopped, I reached into the hole and discovered a large pile of metal ore pieces at the bottom. I scooped up all I could grasp and brought my treasure up into the daylight. I could hardly believe my eyes. I had numerous pieces of ore that I instantly knew were high graded. My hunch had paid off. A miner, decades ago stole pieces of silver, concealed them on himself until he could bring them out of the mine, then buried his treasure for safe keeping. What I discovered was a high-grader's stolen silver cache!

I placed the silver on a large cloth and when I was satisfied (yet with a feeling of bewilderment) I recovered all the silver pieces I picked up my detector and inserted the searchcoil back into the hole (a dedicated treasure hunter always double checks a hole he's dug). I rotated my searchcoil, scanning the hole. The

TOP: Bill Mason, of St. Paul, MN, specialized in searching old abandoned farmhouses. These two caches, are separated into nickels and silver. Perhaps the nickels were only a diversionary cache to satisfy a searcher in hopes the silver would not be found. BOTTOM: A small portion of a 120 lb cache of 200 year old man-made coins found in fireplace ashes in Batopillas, Mexico. There are four distinct sizes, which reflect denominations similar to $1 and .75, .50 and .25 cent pieces. Metal silver is poured into cactus leaf molds and when cooled is chopped into these four sizes by weight. Note the straight edges.

detector sang out again, but not as loudly as the first time. I used "buddy" to jab three or four times in the hole, and each time more silver ore pieces fell from the cache hole down into the large cavity.

By the time I placed all of the silver ore on the cloth, I had amassed a large cache of stolen silver ore. But I remembered my detector had told me there was more metal buried there. I inserted my searchcoil back into the hole and began scanning the sides. Suddenly, at the same depth as my first cache, but on the opposite side of the large hole, my detector pinpointed a second large target.

I thought, "another silver cache! Two in one day!" I jabbed Buddy in the side of the hole exactly where my detector said, "dig". Unbelievable, but true, I heard my second "Las Vegas jackpot!" It was the same unmistakable sound of a second silver dollar jackpot paying off.

I placed "Buddy" back into the hole and jabbed three or four times and each time more silver ore pieces fell into the hole from its decades-old slumber. Upon examining some of the pieces, I discovered this ore was identical in nature to the ore I found just a couple of minutes before.

A double Las Vegas jackpot treasure. I truly could not believe what I was seeing. Earlier, as I began taking the metallic pieces from the first hole, Roy came over, grabbed my camera and took the two photos which you see in this book. (See also Page 70).

I have since analyzed this story and realized that as I sat there eating lunch with Roy, I thought about man's human nature and greed. It is human nature that if you give some people the opportunity to steal something they covet, they will take it. I realized there were probably men, from decades ago, that were mining here and saw the opportunity to steal silver, and they did. Their greed caused them to steal more and more.

Thinking about human nature and greed, I believe, caused me to scan that day around this rock simply because I recognized it as a perfect marker for someone to hide something. Its size is so large it would easily be remembered. I can even close my eyes and visualize how the high grader sat beside the large rock one day at lunch and as he ate, took his knife and dug a hole where he placed the cache of silver. Then at a later time, he sat on the ground and leaned back against the other side of the rock and dug another hole and buried another one of his silver caches for me to find.

So, there you have it. Who cache hunters are and the basics of how they hunt for caches others put down. Keep reading to see how you can join this dedicated group of treasure hunters who pursue those really big prizes still waiting to be discovered.

Chapter Two
Defining a Cache

What exactly is a *cache*? A dictionary defines it as "a hiding place" or "something stored in a secure place." Frontier explorers often cached food and provisions in a secure place so they they would be available for future or emergency use. It is believed that some of the caches of the Lewis and Clark expedition, which explored the Louisiana Purchase between 1803 and 1805 were never used and still await discovery. To find one of these would be historic indeed.

However, we're not talking about food, clothing or other provisional caches. We're talking about treasure lost long ago and abandoned by the hermit farmer who didn't trust financial institutions or the bank robber arrested with empty hands. The possibilities are endless and it's one of these treasure caches, money caches if you will, that we plan to find.

Caches come in every size and shape. Let's discuss some of the sizes of typical money caches that you are likely to encounter.

The first is the buried or hidden coins or bills in a tobacco tin or other small can. Many old cans are still waiting to be found even though such a cache may not be common today since tobacco is no longer sold in those convenient cans. Still, there are other appropriate containers.

Another small cache would be a coin purse, either hidden away in an old house or simply lost. Those are typically the smallest type of cache you'll find. Of course, small coin purses lost by young people are often found in parks.

An example of the next in size is that of the little, old lady who had a secret horde. Maybe she was a farmer's wife who lived over a century ago. Many times, believe me, people such as she had a money cache with coins or bills they reserved for a rainy day. Here she is then, the lady with her stash in a fruit jar that she doesn't want her husband to know about. If she hides it in the house, he's going to find it. Or, some child will stumble across it. Where to cache it?

She has her own private garden spot where she grows flowers or vegetables. So, the next time she goes into her garden she puts the fruit jar in the folds of her dress. She realizes that nobody will know that she buried her treasure in her own garden. Perhaps it's still there, behind the old farmhouse waiting for you to find. The lesson to be learned is to be sure to locate and work the garden area.

What's the next cache size? This will be the farmer's horde. One that he has purposely saved and could be sizable. In many cases there could be two caches. There was also a small horde that he hid in a place where it could be reached quickly and conveniently. It could be in the house or in an outbuilding where it's easily accessible. But it would hold only a few bills or coins.

He could take it to the barn where he cannot be seen, and he'll conceal it. He'll go into a horse's stall and bury it there because he knows the horse's movements back and forth will completely hide all traces of it. Or, he'll bury it under a pile of cow manure. Who's going to dig there?

He might put it in a post hole. Many hobbyists have found one or more of these so-called "post-hole banks." Look for a fence post that appears to be sitting loosely in the ground so that it can be removed easily. It's not holding up the wire, the wire is holding it up! Look under those for the larger caches. See Chapter Four for additional information on the construction of post hold banks.

ABOVE: This treasure hunter, (Center with blue jacket) sat on the back porch of an abandoned farmhouse, eating his lunch and observing the area. He noticed, 30 to 40 feet from the back door, a slight rise in the terrain. He made a sweep across the area and got a loud signal and found broken glass and a coin appeared. He dug more carefully, and found coin after coin. Charles Garrett enthusiastically examines the man's cache while Vaughan Garrett video documents and records the man's success story.

LEFT TO RIGHT : Charles Garrett, James "Monty" Moncrief, Bill Fulleton, Roy Lagal in search of adventure, camaraderie and a treasure cache estimated at over $20 million.

Coins in large jars are popular targets of cache hunters. Early in the last century when banks were not as reliable as they are today, individuals often chose to "bank" their money in such jars.

I participated in the search for several caches including one that once belonged to a doctor who used large jars that once held medicine. He buried these around his home and in his garage. It's interesting that he often buried the jars in the middle of his garage. Thus, when cars were parked there they would cover the site of that particular cache. Anyone who wanted to dig the cache had to crawl under the car, a formidable task.

In one building I discovered a set of numbers on a wood joist. These turned out to be the combination of a safe which supposedly had never been opened. When we opened the safe, however, it was empty. Members of the doctor's family remembered an individual who visited the site several times and left each time with a bundle he placed in the trunk of his car.

It was known that the doctor owned a large quantity of gold coins, but the ones we found were silver. As the story stands, it remains a mystery where the gold is.

Clearly, the loot from a bank robbery or stagecoach holdup is the next largest type of cache that is still being found. The loot could have been held in a box, small safe or saddle bag. It was hidden in whatever was available at the time of the robbery.

Most likely you've seen an Old West movie where the outlaws rob a stagecoach. They throw down the strongbox and the bandit snatches it up, jumps on his horse and rides off into the sunset. However, when the posse eventually catches up with him there's no sign of the strongbox. Where is it? Maybe it's still waiting for you to find!

Now, we come to the largest cache, what I call the family treasure. Often it contains items other than coins or bills. It might contain silverware, priceless relics or anything the family considered valuable enough to save. One such family treasure that was found was a 55 gallon can filled with turn-of-the-century silver dollars.

It's possible that the family who failed to recover it used it to protect another similar cache. This is what I like to call "the misleading treasure" or "the false treasure." Generally this one may be impressive in size, but may not have much value. The reason is that the person who hid his family treasure rather than use a bank understands that people believe his family has a lot of money and know they don't use banks.

"So after I bury my horde and put it down good," he says to himself, "I'm going to put another one in an obvious spot. Then if someone comes snooping around and they find this small cache, maybe they'll be sidetracked into thinking they found our money." Yes, that's what some folks did in the past. They decided to bury a small cache just to sidetrack thieves and bandits. Then, they buried their large treasure a little deeper and put the little one on top or at a nearby location.

This leads to a hunting tip that's discussed more thoroughly in Chapter 5. Always check that hole from which you just removed any treasure. There might be something below the one you found.

But, let's close this chapter discussing more about the "family treasure." It, and the robber's loot, generally constitute the largest cache discovery for any treasure hunter. Both will be large and you can be certain that the family treasure will be well hidden. And if buried, it will be deep. It may be hidden behind false walls or in the secret room of a house. When you're searching an old

ABOVE: While searching for a large cache believed to be hidden near this location, Garrett ensures his metal shovel blade is high enough in the air so that it does not interfere with the detector's operations.

A coin cache container found among "junk" stored in an old garage. Be alert
for any container that can be used to hide treasure. One creative person
once hid his wealth in an automobile transmission case, with the oil still in it.

or deserted house, always look for such places. You may be surprised at the results of your persistence and curiosity.

Of course, be respectful of property even if it is abandoned. Don't tear down old homesteads. Obtain permission and use your metal detector wisely to locate the precise location of the treasure you seek. Old homes are often locations of family caches containing money and other valuables. Residents who lived in such houses hid the treasures, but never recovered them. Why? Perhaps, they died suddenly before telling anyone of their cache. Perhaps, they grew old and simply forgot about it. You can probably imagine a dozen other reasons.

Some of my friends purchased an old home and immediately began to restore it to its 19th Century condition. In the master bathroom was an old iron tub with a cast iron skirt upon which the tub sat. They believe, and I agree, that treasure was hidden under this tub within the hollow space created by the skirt. The hired workman who disconnected the plumbing and moved the tub immediately left the job site and was never seen again. He didn't even return to to pick up his wages. What did he find? As with so many other caches stories, your guess is as good as mine.

The moral of the story is that if you're interested in restoring an old home, pay close attention to possible cache sites.

Here are a few ideas to consider: Caches are also referred to as any hiding place, hide-a-way, hidey hole, secret horde, secret place, recess, nook, cranny, niche hole, fox hole, funk hole, mad money, sanctuary, refuge, stash, cubbyhole, cubby, escape loot, just plain loot, rainy day store, deposit, plant, reserve, reservoir, resource, stockpile, bird-in-hand, sinking fund, ace-in-the-hole, cold storage, depository, supply depot, vault, stock room, glory hole and last, but not least, treasury.

Chapter Three
Researching a Cache

Caches are found by accident on occasion. However, a great deal of the really large ones, are found by using proven research practices. If you've read my other books or heard me lecture, you know how highly I regard research. It's vital to any kind of real success in hunting for treasure with a metal detector. And, in no area of our hobby is research more important than in cache hunting.

Most individuals whom I consider "professional" cache hunters spend a major portion of their time in research. Since proper research may require extensive travel, the costs associated with finding a cache can be considerable before a detector is ever used. Sometimes, cache hunters are required to pay sizable sums to obtain information. Research can consume up to 99 percent of a successful search and recovery mission. Usually individuals pay their own expenses.

Without proper research you'll be as lost as a driver without a map in a strange city. You need a waybill, or directions, to guide you to the location of your cache. Such waybills can come from public and private sources. Be cautious of maps that supposedly lead to a cache or even taking the word of anyone as absolute gospel. You must always find the primary source for yourself. To start at the beginning involves a study of basic research material and sources. You must know what you are looking for, where it is and that it actually exists (or existed at a specific time in the past).

Unfortunately, there is no one-two-three research procedure I can outline. I wish I could. What you must do is define your goals.

Frank Angonna, a friend of the author, found this gold bar while searching for treasure caches throughout Texas and Mexico.

ABOVE: Charles and Eleanor Garrett (L) discuss cache hunting with Idahoan master treasure hunters Charlie and Sue Weaver (R) during their visit to the Garrett Corporate Offices. Charlie Weaver is a professional field test operator for the Company. BELOW: (L to R) Wally Eckert (a full-blood Nez Perce), Roy Lagal and Garrett enjoy some down time while searching for gold at an Old West Idaho camp site.

Every good cache hunter will ask the following questions:

- What am I looking for?
- Does it really exist?
- Where is it?
- Will I have clear title to it if I find it?
- Have others looked for it?
- How do I know they didn't find it?
- What will it cost me to find it?
- Is it worth what it will cost me to find it?
- What type of equipment will I use to search for it?
- Will I search alone or include others?
- How will I split the cache? Do I have a signed agreement?
- Do I understand what I must do to settle with the IRS?

Certainly these questions are rudimentary, yet important to consider when beginning a quest for hidden caches. I want to emphasize that you not search for the "will-o-the-wisp". Or in other words, don't waste time looking for caches unless you are sure, based on solid, personal research that it exists. Use all available sources to track down information about the cache you seek.

Establish your goals. Then believe in your work and your ability to achieve it. Finally, work hard to make it come to full fruition. To repeat a truth, successful cache hunting can often be 99 percent research and one percent recovery.

Don't think of research as sitting in the back room of some dusty, ill-lit library where you must pore over volumes of scarcely legible books, articles and newspapers. Research can be fun. It can become something you enjoy and look forward to as much as you do actually finding the cache. And, who knows, you may discover information that leads to another treasure.

When you become obsessed with cache hunting, you'll continually think about it. You'll scan newspapers and magazines for stories and data about local sites. When you talk with people, especially "old-timers", you'll ask them about events and places and people of the area.

There's an unlimited amount of research and information available to you. The only limits will be those you impose upon yourself. Knowing that everyone has shortcomings you should never rely entirely upon the work of others. When someone is willing to write about a cache, you can be sure that person has abandoned his search for it for one reason or another. You must analyze data with a cautious eye. Failure of the writer to complete the research and recovery could be due to the lack of funds, time or simply interest. But, if a person took time to travel to the supposed site of a cache and to investigate it, that person must have believed in the story.

Never give up on a suspected cache just because you have been told that someone has looked for it and couldn't find it. You don't know who searched or when or with what kind of metal detector. In researching my novel, *The Secret of John Murrell's Vault*, I returned to a location where, just like Gar Starrett, I once found only a deep and empty hole instead of the treasure I expected. As I reinspected the hole, it seemed much deeper and larger than I remembered. Then, it occurred to me that maybe the real treasure had been buried below what had been removed from the hole when I first saw it. Perhaps only a sampling of items were left in a container buried above the main cache to satisfy anyone who might accidentally stumble on the site.

Do not become discouraged if in the early stages of your cache hunt you cannot achieve immediate success. Set goals and strive to achieve them. Success will come if you persist. I advise working on several projects simultaneously. Since research and

50

travel can be expensive, it's good to "double up" on the uses you can make of it. Plus, this "cross thinking" (comparing one location with another and letting one investigation help solve another) will sharpen your mind and help you develop the cross-reference skills that will help you be more successful in all areas of treasure hunting.

Research Sources

Truly all of the sources you have utilized in your other forms of treasure hunting will serve you well as you seek to research caches. Yet, there are some sources that will be more valuable than others. Furthermore, you can access numerous traditional sources on the Internet without ever straying from your computer. Your ingenuity will be your guide here and you may want to investigate other sophisticated research techniques.

- **Libraries** - Here's where you can spend a lot of time that can pay off and provide sources that you can follow up with on the Internet. Unless you're familiar with library cataloging, ask the librarian for help. Tell her you're looking for history books, periodicals, maps and other sources that list local historic events. Take time and look through every reference you find. Either use the copier to gather the information you need, use your laptop or take along some 3x5 cards and list each site on a card. The more specific you can be with the librarian as to what you're looking for, the more help you will get. And, if you are open and friendly with the librarian, you may be told of other individuals who are, or have, worked on the same treasure. I know, this happened to me several times. When I visited the Natchitoches, Louisiana, library and begun asking questions, the librarian immediately knew what I was looking for because I wasn't the first to ask those questions. So, don't be bashful. Ask!

- **Newspapers** - When you have free time, go down to the local newspaper office and browse through old newspapers or microfilm. These are often available on the Internet. Of course, you should always use old newspapers to confirm that events actually occurred and learn how the event itself was described. I get a feeling of having traveled back in time when I locate a news story about the event I'm researching.

- **Old-Timers** - The old-timer is one source of information that you must never pass up. In fact these storage vaults of treasure locations should be actively sought out and quizzed for every last scrap of information that can add to your search for a cache.

- **Park Manager or Proprietors** - One fellow I struck up a conversation with at an abandoned park described the probable location of a safe that was stolen by two men approximately 15 years ago. One of the thieves told him they stole the safe and hauled it across the dam. During their attempt to open the safe they were fearful of being caught by the police so they rolled the safe down the dam's slope into the water. I investigated the area and determined that the dam was soft earth and, in all probability, the safe quickly sank several feet into the earthen dam below water level. I earmarked that one one for future investigation, if I can locate the owner.

- **Old Atlases** - Your library or historical society may have old issues you can review. These will be particularly helpful in learning the names of places as they were called decades or centuries ago.

- **Historical Books** - More and more historical books seem to be appearing and the authors and editors all try to outdo each other. Such volumes can be used to check out and confirm your facts about an event from long ago and its location. I love to research historical treasures and have investigated several caches in foreign countries. There are several I'm working on now, treasure caches in Mexico and WWII arms caches in France.

- **Maps** - Never pass up the opportunity to scan both new and old maps to learn more about your areas of interest.

- **Museums** - Don't be content to just browse through a local historical museum. In guarded terms you can tell the curators what you're looking for; they can dig back into dusty files and perhaps come up with some additional information, especially about the very existence of your cache.

- **Ghost Town Books** - Be on the lookout for books written about ghost towns and old sites in your area.

- **Historical Societies** - If the town or city is large enough, there will be a "home" where the historical society has its headquarters. Not only will the persons on duty probably be well versed in local sites of interest, there may be a library of invaluable maps and books that will give you more information about the cache you seek and the events that led up to its concealment. They may even tell you some stories that can point you to caches other than the one you're seeking.

- **US Forest Service** - The US Forest Service, as well as state forest services, maintain an excellent photo and map library that may contain photos of areas of interest to you. Check with your local Forest Service office or write to:

> Forest Service, USDA.,
> South Building,
> 12th and Independence Avenue, SW
> Washington, DC 20013

- **National Cartographic Information Center** - This center offers 1.5 million maps and charts, 25 million aerial and space photographs and 1.5 million geodetic control points. Write to:

> National Cartographic Information Center
> US Geological Survey
> 507 National Center
> Reston, Virginia 22092

- **US Government Printing Office** - The US Government Printing Office offers more than 25,000 books and pamphlets through a centralized mail order office and 24 bookstores throughout the United States. To have your name added to this free descriptive booklet distribution write:

> The Superintendent of Documents
> US Government Printing Office
> Atten: Mail List - Washington, DC 20401

- **National Weather Service** - Climate data and flooding information could be of interest. For general information write:

> National Weather Service
> National Oceanic and Atmospheric Administration
> 8060 13th Street
> Department of Commerce
> Silver Spring, MD 20910

- **State Archives** - During normal business hours you can search through historical documents, maps, charts and prints relating to the history of just about any state. Since the archives are funded by tax dollars, you certainly shouldn't overlook this source.

- **Treasure Maps** - Most maps of buried treasures are mass produced and sold for $1 to $5. The best use for these is for decorating your family den or study. The "X" indicates a specific treasure location and usually denotes an area of many square miles, the description of each site is vague and the value of each treasure is probably grossly exaggerated.

Other types of charts and maps, however, can be important research tools for today's cache hunter. In fact, before you begin research for a particular cache in a specific location, you should secure as many maps and charts of the area as possible. They can prove as valuable a tool as your trusty one-touch detector, and you may be surprised at the different types of essential information they contain.

The best topographical maps for treasure hunting purposes are those in the "seven-and-one-half-minute" series, with a scale of 1 to 24,000. Each of these maps cover an area of approximately 60 square miles. These maps can be obtained from the U. S. Geological Survey. Information on them can be obtained at any Government bookstore or by writing the Superintendent of Documents, Government Printing Office, Washington, DC. You should request information about all maps of areas you are interested in such as geology, transportation, historic sites, mines and abandoned settlements.

The Bureau of Land Management, Department of the Interior, Washington, DC, sells a large series of charts called "Public Land

ABOVE: This lovely, very old ghost town was truly a joy to walk around while thinking about the human interest stories that took place there during the early decades of the 20th Century. The peaceful nature of this place is a joy even to just think about. BELOW: The author hands a VLF Grand Master Hunter down to his brother, Don Garrett. Don searched the cellar, which remained unsearched for decades.

Maps," which are updated periodically. They are invaluable for determining whether or not your potential site is situated on private land or public domain, especially valuable information for persons prospecting gems or minerals. The United States Forest Service and the National Park Service also publish and distribute maps showing all lands under their jurisdiction.

Many states also have geological survey departments and other agencies that produce and distribute excellent maps that have proven to be invaluable for treasure hunters. Some states have maps dating back more than 100 years showing the exact locations of mines and mining settlements. In fact, many local chambers of commerce, tourist bureaus and historical societies sell or give away maps showing historical sites in their areas. Regular road maps usually show the locations of historical sites and in many cases even abandoned settlements.

Another method of locating abandoned settlements is to obtain access to copies of all old maps of a given area. By matching them to a modern chart you will notice that certain places no longer appear on the modern maps. Such places may have become ghost towns.

Maps made by the Earth Resources Observation Satellite can be acquired from the EROS Data Center in Sioux Falls, SD. Designate longitude and latitude of four corner points to denote the area whose satellite map you want. You can select just one point, however, for a 10-by-10-mile bracketing.

Speaking of maps, when people learn you're a cache hunter, you'll probably be offered the "opportunity" to buy the map leading to a sure-fire treasure. Well, such treasure cache maps are generally worth about a dime a dozen. But, you'll probably be told an exotic story and be asked to pay a lot more than that. My advice is: Don't! If the map is authentic, why is someone trying to sell it to you? Why isn't he or she going after the cache itself. Beware of any "treasure map" unless you know something of its authenticity.

After saying all that, I want to tell you a story, which begins in Juarez, Mexico, about a map found in an antique chest, obviously very old and hand-made. On examining the chest the owners discovered what looked like an animal hide tacked on the bottom of one of its drawers. The skin had no marks on it, but the couple suspected something and took it to my friend, world famous treasure hunter Bill Mahan.

Bill put the skin under ultra-violet light, and a line drawing of a map with Spanish wording appeared on it. You'll find the rough outline of the map printed in this book, but I've removed some details for obvious reasons. Simply stated, this is a map of a village located on a plateau overlooking a river. The inscription on the map suggests that there are seven tons of gold and/or silver buried at a secret cache site.

Bill Mahan searched for this treasure for several years. Shortly before his death, he gave the map to me. Bill located one site indicated on the map. I've learned of nearby abandoned mines that once produced gold and silver. As I investigated, I learned that a gold bar was found nearby.

I'll tell you that I have searched many times for this treasure and haven't found it...yet. I believe it is buried in Mexico. I believe its burying place was brillantly conceived, it's self-guarded.

There is no way that I can stress strongly enough the importance of all the ideas presented in this and the following chapters. Sure, there are caches just waiting to found, and I am sure that those of you who persevere will find your own as a result of your own research, investigation and hard work.

I hope that you will understand, however, that considerable effort will be required if you are to discover a cache.

If you become a dedicated cache hunter, you'll be faced with so many opportunities that you'll never be able to pursue all of them. I know, it's happened to me.

I remember an old house that my brother Don and I investigated shortly before it was to be destroyed. It was a mess. Apparently, others rifled through it looking for valuables with little thought of neatness. In fact, the last occupant of the house had littered the place with empty dog food cans.

However, we found hundreds of old magazines, as well as a ledger with the words "reward for the return of this book" written boldly on the cover. Wondering what could have made the book so valuable, I studied it and found that it contained voluminous records of the man's rifle marksmanship. Near the end of the ledger, however, was a scribbled paragraph containing words such as theater and telephone pole, along with distances and directions. Interspersed throughout the paragraph were numerous curse words. A waybill to treasure, I presumed.

My brother and I left with the magazines and a few other items and I promptly forgot about the ledger. All of this occurred many years ago when I was building our metal detector company and had little time left over to hunt for caches.

Months later, I remembered the ledger containing the treasure waybill and decided that the "rifle scores" may have been a tally of the gentleman's buried coins and other treasures. I looked through the boxes of magazines that we took from the old house but was unable to find the ledger. Did I leave it behind?

Another missed opportunity. But one to dream about.

Chapter Four
Locating a Cache

Where are caches hidden? They are hidden in as many ways and in as many places as it is possible to hide something valuable. Caches are found everywhere. I break down caches into two major categories:

- Those hidden by an individual after careful planning
- Those buried in haste

Of course, these two categories refer basically to the wealthy farmer who hid his family treasure and the bandit who hid his loot because he knows he's about to be captured.

Generally, you're going to find carefully hidden caches in places that were familiar to the people who hid them so that they could remember where they cached their wealth. Because a vast majority of people tend to forget, the person burying a cache will use a marking and numbering system with which he or she is familiar.

A treasure hastily concealed is going to be found in the best location that existed at the time the cache was hidden. When you hear a story, or get a treasure map noting the location of a cache buried on a mountain or in other difficult-to-reach locations you should ask yourself, "Why there? Would someone have scaled a high mountain or climbed down a steep ravine to bury a cache?

Always consider the reasoning of someone who has buried a cache and you'll improve your odds of finding it. It won't be just "luck," either! Whenever you're tempted to attribute the success of another cache hunter to "luck," remember what the football coach said when they accused his team of being lucky: "We had to be there for the luck to happen!"

Here are some of the more common places where caches have been found:

- **Ground** - Most caches are found in the ground, which is about the only place to put them in many cases.

- **Rock Slides** - In the beautiful northwestern United States, stories abound of caches found in rock slides. It was a common practice for 19th Century folks to hide their wealth here.

- **Wells** - This location is frequently overlooked when searching for hidden caches. Here are two stories of caches being found that I know about or have heard told. An old iron steam kettle was filled with silver dollars and was lowered down into a well with wire that was tied off. Another was discovered in a cistern. When it was built, a ledge was located a short distance below the opening where it widened out. It was on this ledge that the money cache was found. So don't overlook wells, cistern walls and other man-made water storage structures. But, be careful. Such areas can be dangerous to work around. Cave-ins do happen.

- **Fireplaces** - These are very common locations for caches because it is easy to construct secret chambers. Such chambers are usually concealed with a lose brick or large rock. Caches aren't normally found inside where the heat is, except down in the ground. One cache I can speak of contained 115 pounds of handmade silver coins and was found in the bottom of a fireplace. If you scan over the floor of a fireplace and get a signal, it could be a tin can or something else. But, don't let anything like that throw you off. Also, make certain that your signal doesn't come from some metal outside of the fireplace or the brick support. You don't want to tear down a fireplace because an iron support rod gave you a detection signal.

- **Base boards** - A typical location is behind room base boards simply because base boards can be removed. Look and you'll see short length boards that can be easily removed. Loose baseboards can be found above doors or any place where sections of molding can be easily removed. A friend of mine found a nice cache above a door behind the framing board. It contained a watch, a knife and some valuable old coins.

Remember, a person who hides a cache in a fireplace or behind a loose base board is *always* looking for a place that lets him recover his wealth quickly and easily. And keep in mind, smaller caches are often placed in an easy location to find in order to protect larger caches from being discovered.

- **Motels / Hotels** - Some years ago my colleague picked up a motel room's telephone book and a purse fell out of it. There was a considerable sum of money and no identification. The point is to look everywhere in motel rooms, under furniture and in shelves and drawers. Do not, however, tear the room apart. Simply look for the places someone might hide or lose valuable items. I've found numerous valuables that were left behind in hotel and motel rooms. Recently, my wife and I were staying in a San Antonio, Texas motel. A man was screaming at the clerk because he claimed that someone had stolen $1200 he hid in his room.

- **Books** - Always look through any books that you find in deserted houses. Money may be hidden in them. Gold and rare coins are often hidden in the spines of old books. Be sure to scan them with your detector. You'll be surprised how quickly you can search an entire library.

As discussed in the previous chapter, I found a ledger in an old house. On the first page was written, "Reward for return of this book if lost." I flipped through the ledger and found a waybill to treasure written on one of the pages.

Take note that every story in this book is true. I tell it like it is, or was. Believe it. Sure, some information treasure seekers hear may be stretched somewhat. But, do as I do and check the validity of each story. You won't go wrong if you do.

14"

ABOVE: This Spanish bell, which dates back to 1410, was purchased by Javiere Castellanos who gave it to Charles Garrett during his expedition to Central America. The bell is currently on display at the Garrett Treasure Museum located in Garland, Texas. Note the cross above the bleeding heart located above the 1410 date.

- **False rooms** - I can tell you that a valuable, antique roulette wheel that was found in a false room in an old gambling hall in Murray, Idaho.

- **Trees** - Money may not grow on trees, but it sure does get hidden in them. Be sure to search the forks of trees. Let me tell you a story of my own experience with one. Several friends and I were deep into Mexico when a local resident began talking to us. He noticed a digging tool that A.M. Van Fossen had on his belt and asked, "What will you take for that?" Van replied, "I don't know, what are you offering?" His answer, a cache of silver. When Van told him we were interested, he nodded and walked over and pointed to a tree and to our metal detectors. When Van scanned the tree, the signals nearly blew the earphones out. Someone, sometime in the past had cached a horde of silver in the tree, which grew around it.

Let me tell you another cache story. On my first trip far down into Central America in the 70's when I was testing the prototype of my discriminating detector, we were staying with a native family. People were coming from all over their remote village to see this strange contrivance that could find treasure. We were demonstrating the new metal detector, showing how it would tell the difference between iron and a coin. One particular man simply stood back, listened and watched. At about midnight he said, "I know where there is a cave. In this cave there are two iron chests. One is filled with silver coins, another is filled with gold coins. There is a solid gold statue of Christ, relics, bells, all kinds of things in the cave. They are ancient."

When we said we didn't believe his story, he became angry and shouted, "I'll prove it," as he stormed off. At about dawn he returned with a large bell that's now in our Garrett museum in

64

ABOVE: Professional treasure researcher and author, Karl von Mueller, and Charles Garrett enjoy telling treasure tales in 1972 during an annual California treasure meet. The two became lifelong friends after meeting one another around 1970 when the Garrett family visited Karl and his wife in Segundo, Colorado. Karl encouraged Charles to write for the treasure hunting industry. As a result of that first meeting, Charles and Eleanor founded the RAM Publishing Co. and began by first publishing Karl's book, *Treasure Hunters Manual #7* and Charles' first book, *Successful Coin Hunting*, which to this day remains the industry's most important coin hunting guide.

Karl and Charles remained a strong treasure hunting team until Karl's death nearly two decades later. Karl is well known in professional circles because of his work on the much heralded and sought after, Spanish LUE caches.

INSET: A photograph of a special, miniature metal detector Charles Garrett created especially for Karl von Mueller and his friends.

An early farmer's cache, such as this one discovered in Oklahoma, are still being found in abundance across the United States.

Garland, Texas. Cast into the rim of this bell is the date 1410 with another date, 1553, hand chiseled into the decorative band. He said, "This came from the cave." Javiere Castellanos bought the bell from the man and gave it to me.

I researched historical data and learned that during the mid-16th century when local Indians revolted, mission padres cached their valuables in a cave. Our friend found that cave, but would never tell us where it was. So, all we have to show for this story is the bell in our museum. One last comment to stifle your curiosity: the man confided that he would never reveal the location nor remove the treasure because he believed that when Mexican authorities found him out they "would shoot me, throw my body into a ravine and cover it with rocks." Another missed opportunity, but I'll tell you more cave stories later.

When I first hunted with a metal detector in Greece 25 years ago, results were consistently outstanding. We discovered coins thousands of years old and occasionally bronze figurines and broken pieces of statuary.

While hunting on the site of an ancient city, I received a strong signal and recovered a very old copper coin. Following my own rules I checked the hole and dug up an identical coin. I continued scanning and digging and recovering coins from a very small area about two feet in diameter. I had obviously discovered a cache of coins that had been buried in an earthen vessel of some sort. Repeated plowing over the centuries had broken the vessel and scattered the coins.

Through research we learned that the coins dated back to circa 400 BC, which made them the oldest coins ever discovered with a metal detector. Since then, however, older coins have been discovered. I still have all these coins, and some are on display at our Garrett Museum.

- **Driveways** - My wife and I visited the late treasure hunter, Karl von Mueller, and his wife for several hours one afternoon. As we visited he said "Charles, do you know where my treasure is?" I said "No, Karl, I don't know where you hide your money." He said "I didn't say money; I said treasure. Come on, I'll show you". He took me out to his driveway and he said "Look around, Charles, where is it? Do you see it?" Well, I didn't see it. He said "It's in the driveway, in the gravel!" He had mixed a lot of silver nuggets and ore with gravel and spread it out on the driveway. Who's going to know? I have heard of silver dollars mixed in concrete and put down in concrete slabs.

- **Paths** - Search paths that lead from the house to its outhouses. These areas are familiar, and a location here would be easy to describe so that another person could locate it. Here's where you might find that housewife's bread and butter cache that she buried many years ago; or, better yet, you'll probably find her treasure in a garden plot. More on this woman's "rainy day" cache later on.

A treasure that I've sought for many years stems from the story of an East Texas farmer dying of pneumonia who had to be restrained by family members as he kept trying to get out of bed. In desperation he said, "I need to go down the lane and get my money." Unfortunately, he never made it to his money. Since the farmer's house was located at the intersection of two dirt roads, there were four lanes from which to choose. What direction to travel?

I have searched along these roads for this treasure without success. Why has it been so difficult? Perhaps it's a misinterpretation of what the dying man said. But several people,
68

Idahoan, Bill Fulleton, searches for an abandoned treasure cache, which is believed to be hidden within the foundation support timbers of this house.

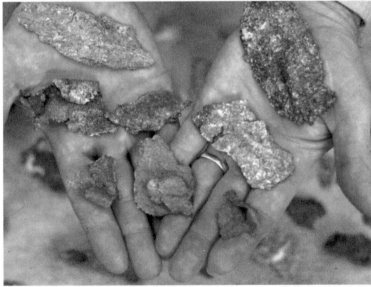

Garrett with two caches of silver ore he found in Cobalt, Ontario, Canada, while searching near a mine. Found near the base of a large boulder, Garrett believes these two caches were high graded from the nearby mines by a miner who worked the mines sometime after 1910. Read more about this "Double Las Vegas Jackpot" cache in Chapter One.

including my father-in-law, heard him say, "I need to go down the lane and get my money." Unfortunately, I did not arrive on the scene until 10 years after the farmer's death. The house had long since disappeared; only traces of the outbuildings were left; bushes had been allowed to grow wild. So, many of the landmarks were no longer there to guide me. And, too, I was not familiar with the old gentleman's habits.

Keep one thought in mind. Chances are that the person hiding a cache buried his money in sight of his bedroom window. Thus, if he heard his dog bark or some other strange noise in the night, he could just look out a window to see if someone was digging for his treasure.

A friend, Edgar McClain, once asked me to search for such a cache buried by his grandfather outside his bedroom window. Edgar died before we located the treasure, and I never continued the search. Someday, I'll recover it and give the treasure to Edgar's wife and two sons.

- **Outbuildings** - A likely burying place might be in an outbuilding such as a chicken coop, barn or any others with dirt floors or where floor boards could easily be removed. I know a man who successfully searched a chicken coop in the dead of night during a violent thunderstorm. Think about it, and you'll be able to come up with the reason for the man's peculiar search methods. The fourth most likely place would be, as discussed earlier, under a fence post.

Fence Row Search

If you don't have your detector with you or if you just want to make a quick fence row survey, walk slowly along the fence and observe the staples holding fencing wire to the posts. Be

bottom to enable the fencing wire to be removed easily so that the post could be lifted from the ground for quick access to a "post hole bank."

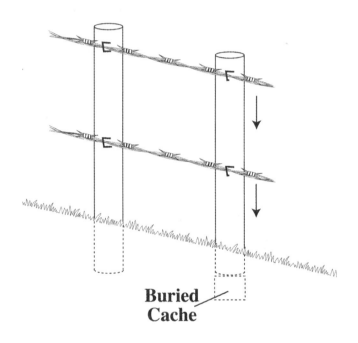

Buried Cache

Search under trees. Once again, the surroundings are both familiar and easy to remember. Keep your eye open for spikes and nails in overhanging branches and search the ground area directly beneath such spikes. Caches have been found at these markers. As the tree grows any spike that may have been nailed into it will remain in the same position. It will not change except in areas of strong prevailing winds. Cache sites are marked by driving a spike or nail into a tree limb and tying a string to it. On the end of the string tie a small object and let it swing as a pendulum close to the ground. Caches are commonly buried directly beneath the "pendulum." This makes it easy to find the cache by hanging string on the nail, weighing it down and digging directly below the weight.

ABOVE: The Garrett Infinium was designed essentially for one purpose, to provide the detectorist with an instrument that could search deeper than any other detector over the world's toughest ground mineralization, including underwater sites. While these heavily concentrated mineral areas are rare, the minerals can cause virtually every metal detector to loose depth. Some land areas in Australia, the United States, Italy and a few other areas, contain very sparse (and usually isolated) areas of these minerals. The Infinium LS has been proven in the above areas and others to operate virtually flawlessly wherever needed especially in these tough-to hunt mineral zones. However, in all other mineralized areas, which are most commonly found around the world the GTI 2500 is the professional's choice for all-purpose treasure hunting.

Scan in and around all outbuildings, especially chicken coops, under dog houses, bee hives, garages and carriage sheds. My friend, retired Special Agent FBI, David Loveless, advised team members to search several places around a ranch where it was believed a cache of weapons was buried. They found the weapons beneath one of Dave's suggested places, a manure pile!

Get into the habit of scanning with your detector as you walk a straight line between the corner of a house and an old well. Scan a straight line between two trees. Scan the area in the exact center of a triangle formed by the position of three trees. People preferred to bury their caches in locations such as these because it gave them a quick reference when they wished to make a recovery.

My colleague, Roy Lagal, has proved to me time and again the importance of these "reference points." When seeking a buried cache believed to be buried in a yard, Roy will draw a location map that includes all buildings, trees, wells and other such objects. Then, he draws straight lines between the various structures and studies their relationship, especially where they cross. By scanning along these lines Roy is remarkably successful in finding caches.

Pick a spot where you can sit and observe the "lay of the land." Notice where the ground rises, where there are obvious mounds and places for you to search. These would include such locations as walkways leading into the woods or passing through gates.

A customer visited the Garrett factory and told us of a coin cache he found buried in the center of an earthen mound in the back yard of a farmhouse. The earth had crushed the glass jar, but the coins were in excellent condition.

74

When searching a farmyard for a money cache, look closely at specific objects and obstacles in that yard, such as a well, the corners of the farmhouse and its chimney. Search inside the chimney and all outbuildings, especially those that contained animals with loud vocal cords.

Never fail to search an old garden area. Here's where the farmer's wife may have hidden some "rainy day" savings in a fruit jar. Remember that when people buried caches, they didn't want to be observed. It would be quite normal for a farm wife to hide a jar of money in her apron, carry it to some special location in the garden and "plant" it secretly. Rings are also found in old gardens. Baby rings are sometimes found in sandboxes.

A location from the 1940s that's still on my "active list" is an area where a blacksmith allegedly hid a coin cache. I searched this area 25 years ago and was forced to give up because of the large amount of trash metal at the site. Today, however, I plan to use my GTI 2500 imaging detector. The small pieces of metal I'll leave in the ground. I'll only dig the larger ones.

Here's another method I will try at the blacksmith's cache site: I'll use the two-box Depth Multiplier searchcoil. As the discussion of this device in the following chapter will point out, small objects will be ignored. Only large objects will be detected.

Plus, here's another thought: I suspect that the blacksmith wasn't content just to bury his cache. I imagine that after he placed the cache in a deep hole, he placed several large iron pieces on top of it. Why? Well, if I buried a valuable cache, I'd try to camouflage it in some way and I suspect that's what he did.

You're probably asking if that iron junk will "mask" my cache target. Remember, that since the Depth Multiplier is used only in an All Metal mode, it will detect all types of metal. The camouflage

A photograph of a very old animal hide treasure map hidden in an antique
Spanish chest. The markings can only be seen under ultra-violet light. The
skin measures 10"x14"

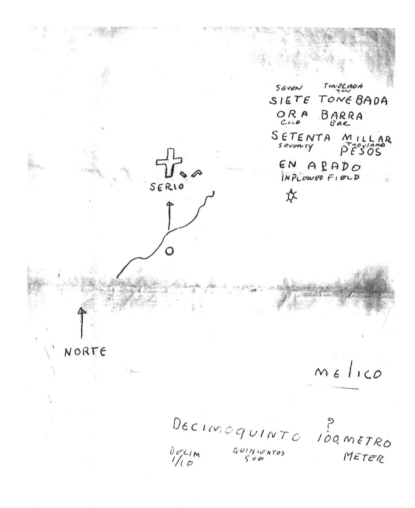

An actual copy of the map found in the old Spanish chest. NOTE: Several important details have been removed from the map. Some of the Spanish words have been translated into English. The author has precisely located the turtle and river locations.

therefore actually adds to the metallic content of the cache, thus improving the Depth Multiplier's chances for success.

A Texas / Mexico Treasure Cache Story
As I mentioned earlier, when friends of famed treasure-hunter Bill Mahan purchased an old chest, they had no idea a waybill for millions of dollars in gold and silver would be tacked to the underside of the bottom drawer. But that's just what happened.
While cleaning the chest, they discovered an unusual looking animal skin. Although no markings could be discerned on the skin, they asked Bill to take a look at it. Upon examining the skin under an ultraviolet lamp, Bill discovered a drawing that revealed several lines, Spanish wording and the name of a Mexican village located somewhere in Texas or Mexico. Even though Bill's friends were excited about the possibility of finding such a treasure, they graciously relinquished the unusual treasure map to their capable friend.

Bill anxiously began his quest for the hidden cache of gold bars and silver pesos he knew were waiting for some persistent treasure hunter. He searched tirelessly throughout Texas during the next decade for his sure fortune. Sadly, Bill would never achieve this dream. His race would end when terminal cancer took over his wish of finding the treasure. Before completely succumbing to illness, Bill turned the map over to me. We spent hours reviewing all of his previous fieldwork as well as the map's symbolization. Of all the information gathered over the ten years nothing suggested that the map was fake. In fact, an abandoned gold mine was discovered in the vicinity and a gold bar was found near a river crossing at the center of the map's believed location.

To this day, I continue the unwavering search for the fortune, destined to be one of the biggest in the history of treasure hunting. The thought of seven tons of gold bars and seventy thousand silver pesos is certainly worth the years of faithful searching, and I have already decided that if I do not locate the treasure in my lifetime, my children will take up the search.

Please note that several of the map's important words have been deleted, however the lines and symbols are relatively unchanged. Even though several Spanish words were misspelled when the map was drawn, I believe I have stood in the exact spot where the person who drew the map stood when it was drawn. Note the amphibious turtle symbol, which I discovered at the site plays an important part in the deciphering of Spanish treasure maps. Cache hunters have learned that many other symbols play an important part in treasure and way bill encoding.

Chapter Five
Recovering a Cache

Okay, you've done your research. You have proof that your cache exists (or, at least existed at one time). You may even believe that you know precisely where it's hidden. Now, exactly how do you use your metal detector to find it?

Remember, I said that cache hunting is different. Well, metal detector techniques necessary for successful cache hunting differ somewhat from those used in day-to-day treasure hunting.

I suggested detectors that cannot search in an All Metal ground elimination mode should not be expected to find a deep cache. To find caches most effectively, I recommend you use the All Metal ground elimination mode almost exclusively, simply because caches are usually buried deeper than most other buried treasure. And, most true All Metal modes will detect much deeper than any Discriminate mode, even when those Discriminate modes are set to "zero." I'll talk more about this later in this chapter.

Let's say you're looking for an iron pot filled with coins. In such a situation your metal detector will signal the presence of the pot, not all those coins inside it, unless the iron container is completely rusted out. Then you'd probably detect the coins. Remember, not all caches are large. In fact, the vast majority are small and buried deeper than single coins are found. Thus, either way, you'll need all the detecting power you can get. Use the All Metal mode, but remember what you've learned earlier. If you seek *real* success, you must be prepared to dig your share of junk!

If, however, you grow weary of digging iron junk, you may be tempted to use some discrimination to reject those small time-

Garrett scans the walls of wooden structures using small searchcoils because the width of the wall and chimney is minimal, hearthstones and all outside areas should be scanned with a deep seeking detector with a 12 inch or larger searchcoil. NOTE: Headphones should always be worn as they provide improved depth and discrimination of target detection signals.

Professional treasure hunter, George Mrockowski, shows several Wells Fargo silver pieces he found during his many years hunting in the desert areas of Yuma, Arizona and Nevada.

consuming nail and bottle cap sized junk items. Remember, that in the Discriminate mode iron tends to "cancel" somewhat the valuable treasure signals. This means that your detector might ignore very small caches and individual coins.

I remember a competitor who advertised, "I find the little treasures." I believe he used this theme because so many hobbyists despair over ever finding a big one. Still, I know that this competitor was successful in finding a number of very large caches. All that is required is effort in research and in recovery. How would you like to run your hands deep down into a large box filled with old Spanish coins?

Let me stress, however, that you should never overlook the search for individual coins. Cache hunters occasionally find them scattered near the burial site of a cache. Several reasons can be given for this. Caches are usually hidden at night when coins might be dropped and overlooked. Also, any retrieval of all or part of a cache would probably occur at night. Again, coins might be dropped and overlooked. Quite frequently digging tools lost by the person who buried the cache can also be found that will help mark the location of a cache site.

Another factor to consider is that many caches are buried in glass jars which can break because of soil pressure or when struck by a plow. Coins will be scattered by additional plowing and the burrowing of animals. I know of one cache of a thousand coins that were found scattered over a five-acre field while the bulk of the cache was concentrated in a ten-foot-square area.

Yet, if you use a detector that offers only the Discriminate mode, even when set at zero, you want to be certain how much residual discrimination your detector furnishes at this setting. Most detectors will provide *some* discrimination, even at a zero setting. Such a small amount of discrimination is designed to avoid detection of iron rust, nails and other small bits of ferrous trash.

When searching for caches, always try to use detectors that employ the latest technology. It is truly amazing how much more effective today's modern instruments are than those with which we were so well satisfied just a few years ago.

A recent technological advancement in metal detectors that gives the cache hunter a great advantage is Garrett's Graphic Target Imaging. This advanced circuitry measures, with extreme accuracy, the size of all detected targets and visually reports this information to the detector operator. Why is this so important to the cache hunter? Because it lets the hunter dig only the size targets he or she is interested in digging.

Earlier in this chapter I told you that I recommend that you use no target discrimination. Professional cache hunters will take no chances that a small, iron object might cause the detector to give a misleading signal. So, the successful cache hunter has learned to dig lots of trash.

But, now with size imaging, small targets need not be dug, thus saving much valuable time for more productive searching. You simply dig only cache sized targets when you are using Garrett's 12-inch diameter Graphic Target Imaging deep seeking searchcoil.

Of course, nothing is perfect. I guess you will dig occasional horse shoes and large farmer's plow points. You won't however, spend valuable time digging worthless tin and aluminum cans and other worthless junk.

When you're searching for a cache in a building, even though you know that it cannot possibly be too far away from the bottom of your searchcoil, always use the largest searchcoil possible.

84

TOP: Not only is there beauty in exploring mountainous regions, often rock slides such as this provide a happy surprise, hidden treasures. It is easy to remove a few rocks, place wealth in the hole and cover it up again. Here the author demonstrates how easy it is to explore these sites with a deepseeking instrument with ground cancelling abilities. You must use detectors that have the ability to cancel (ignore) even the toughest iron mineralization. The serious cache hunter should use either the Infinium LS or GTI 2500 because of their proven deepseeking and heavy mineralization cancellation abilities.

BOTTOM: Garrett searching for an outlaw cache in a mountain slope rock slide believed to have been hidden in 1895 and worth several million dollars. The Depth Multiplier accessory he is using is an excellent accessory for any cache hunter. It detects large objects at great depths and is not hampered by small metal objects, nor most iron minerals and water.

Chinese coins recovered from the rock wall of an old, deserted house in Southern California. The discovery of this cache can be seen in Garrett's treasure video *Southwestern Treasures*.

Regardless of the size cache you seek, you must not take a chance. So, use a large searchcoil. There is no doubt that even the best treasure hunters have missed deep caches that were beyond the range of the finest detectors available in earlier years. These caches await you and other hunters with the 21st- century instruments capable of finding them.

Remember, remember, remember that larger searchcoils can detect larger objects at greater depths. Money caches have been found at all depths (arm's length seems to be popular), but you want to be prepared for extremes. For example, caches originally buried near ground level will be found at greater depths where washing and drainage redesign the landscape.

All the more reason to use the larger searchcoils – even the Depth Multiplier Treasure Hound! It's easy to use Garrett's Depth Multiplier attachments with a compatible GTI or CX detector. Just press the Power touchpad, and you are ready to go. No other adjustments are necessary. Wear headphones and set your audio threshold for faint sound.

I've long stated that successful cache hunting requires considerable experience and thinking. You must learn to put yourself in the shoes of the person who hid that cache for which you are searching. It's easy to understand why a person wouldn't just run out into his yard haphazardly and dig a hole to bury a jar full of gold coins. No, if you were burying a cache, you'd select a secret place and a secret time to bury it, perhaps, at night. And, your "secret place" would be one that you could find in a hurry while others would overlook it!

Practice this yourself. Put some money (or something similarly valuable) in a jar. Go outside your house and bury it. That's right. Go ahead and bury it, if only for a few minutes. After you've done this, you'll be able to ask yourself the questions that probably

ABOVE: Garrett demonstrates how to use a 4' metal probe, which he occasionally uses to pinpoint caches and other buried objects. To make it easier to insert the rod into the ground, a bullet shaped probe point, slightly larger in diameter than the rod is welded to the probes's end. BELOW: Garrett displays a specially designed hand drill which consists of five, 3 foot extensions giving him an exploring depth of about 15 feet plus the bit length. INSET PHOTOS: Two bits (for wood and earth materials). A: Earth Auger; B: Wood Drill Bit; C: Drilling platform plywood with hole.

occurred to that person who hid any cache you ever seek.

Would you do it in broad daylight? Would you just walk out into the yard and start digging? Probably not, because you wouldn't want anyone to see what you were doing. So, choose the right time and the right place to bury your cache. Can you find it easily? Can it be found accidentally by a stranger? Will it be safe? Many other questions will come into your mind as you recover your own cache and relocate it a time or two. This is good experience that will make you a better cache hunter.

Searching Indoors
When searching for a money cache behind or inside a wall in a house, you can generally use either the All Metal or Discriminate mode and a twelve-inch search coil. But, even when using the Discriminate mode, I recommend that you turn the discrimination controls to their lowest settings, just enough discrimination to eliminate nails from detection.

In either mode you'll have more than enough sensitivity to detect almost any size cache in all walls, despite their thickness or type of construction.

When your treasure waybill leads you to a stucco wall containing a wire mesh, here are some tips to help you detect through that mesh. Place your searchcoil against the wall, set your detector in its Discriminate mode to the lowest discrimination setting (iron or nails). By carefully sliding the searchcoil across the wall mesh interference will be reduced. You may hear a jumbled mass of sound, but you should listen for significant changes that could indicate you have located your large cache. To further reduce the wire mesh sound, you may reduce the audio threshold slightly down into the quiet zone.

Some prefer to search walls with a mesh by holding the

ABOVE: A.M. van Fossen (Houston, Texas) a renowned treasure hunter, sent these photos of a gold coin treasure he found with a Garrett detector. He is famous for his international research, found treasures and black hats. Written in the photo are the words, "I love Charlie Garrett, JJ, implying: "Treasure courtesy of Jesse James!" The arrow points to one of Charles Garrett's detectors used to find the cache.

searchcoil several inches or even a foot away from the wall.

Positioning the searchcoil this far away should take care of the jumbled sound, yet still let your detector detect large masses of metal such as a money cache. If you employ this technique, use a 12-inch searchcoil and practice detecting cache sized objects at various distances from these "target" objects.

Selecting the Correct Searchcoil
The failure of a treasure hunter to locate a given cache can often be traced to the size of the searchcoil being used. There is another way to say it. Use the proper size searchcoil and your chances of finding the cache you seek will vastly improve. Let's delve into the selection of searchcoils to use during any search for treasure caches.

Even though I have discussed searchcoil features in this book, I think it's worthwhile to discuss, in greater detail, the selection of cache hunting searchcoils. First, to prepare yourself, learn as much as possible about the cache you will be trying to locate. Is it buried in the ground, building wall, a rock wall or somewhere else? Is it of a metallic nature or is it paper money contained in a non-metallic container? If this is the case, you can forget about detecting it, you can't!

How deeply do you think it is buried? Has soil erosion caused it to become more deeply buried? Your best bet is to believe it to be much deeper than originally buried. In short, learn as much about your cache as possible.

I will be brief, in my searchcoil recommendations. The more information I give, the greater might be your confusion. When I search for caches, I virtually *never* use a searchcoil smaller than a 12" size. A 12" searchcoil can locate small and large caches provided the depth of the large cache is not great.

I have said that when searching wooden walls you will probably be safe to search using an 8", 9", 10" or 12" searchcoil. But, to add to that statement, you will be the safest, if you use a 12". I believe my recommendation is correct even if your cache consists of only a few coins buried in a non-metallic container such as a leather purse. In fact, many *coin* hunters always use a ten or twelve inch searchcoil. They have learned to discern the smaller detection signals that might occur on deeply buried single coins.

When ground searching, to be safe, use a 12" or larger diameter searchcoil if you believe the cache is buried no deeper than about 18 inches. Deeper than that, use a two box detector. Garrett's "two box detector" is the Treasure Hound Depth Multiplier accessory when attached to any appropriate Garrett model.

If your cache is smaller than approximately an 18-ounce food can, do not use a two box detector, use a 12" searchcoil.

You'll enjoy several advantages when using a two box detector. The first is that you will not detect objects somewhat smaller than a 12-ounce drink can. You will save needless digging of small metal objects when searching around home and ghost town sites. Searching with a two box will also permit you to scan an area faster because the search path is wider.

Remember, select a high quality instrument that the manufacturer recommends for cache hunting. Be aware that there are some difficult ground minerals you may someday search over. To achieve maximum performance from your detector, it must be capable of ground elimination (balancing) over tough minerals. You must be able to manually do the ground balancing or let the automatic feature do it for you. Believe me, if your detector cannot correctly perform over most ground minerals, cache detection will be hindered and greatly so over extremely high mineralization levels. Learn to use your detector to the maximum efficiency you

ABOVE: Of course, the author's favorite instrument is the GTI 2500 equipped with his choice of coils or the Depth Multiplier (as shown above). Charles finds great joy in the instrument's deep seeking abilities and near perfect target sizing ability. He has striven diligently since the day he first built his own equipment in 1964, to use only the best possible equipment. He claims this detector is the summation and culmination of that forty-plus years of engineering and treasure hunting effort.

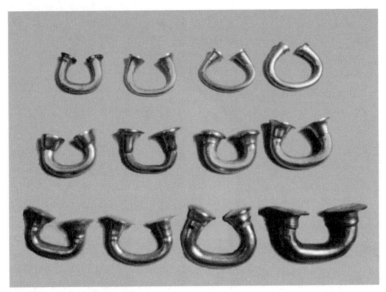

ABOVE: A cache of gold nose rings discovered in Africa. Made of alloys and differing percentages of gold, these rings hold a high historical and intrinsic value and well worth the effort to recover.

and your detector are capable of during all operating conditions.

Always use optimum ground scanning techniques and headphones. Be patient and scan at a moderate searchcoil speed. Think success. When you have learned to achieve maximum scanning efficiency, maximum success will be achieved.

To summarize, follow these searchcoil selection guidelines:

- If you believe your cache is smaller than a 12-ounce drink can, use a twelve-inch searchcoil, and be prepared to dig lots of small junk. The target sizing capability of Garrett's GTI Imaging detectors will, however, eliminate your digging of small junk targets.

- If you don't find your cache, use a two box instrument. Depending upon target size and depth, the two box will give you the extra edge you might need for success.

- If you believe your cache is larger than a 12-ounce drink can, start with a two box instrument. Be thorough in your search and you can find it if it is there.

Now, remember, if you still don't find your cache, you still learned something important. YOUR CACHE IS NOT THERE! Either it was never there, someone else has already found it or you are looking in the wrong place. Don't give up without one last effort.

Review your research data. Prove to yourself once more that your research is complete and accurate. But, don't give up cache hunting. There are still treasures you can find!

Recovery

Most hobbyists don't get involved with caches that require a bulldozer or other kind of power shovel. So, sturdy digging implements are the primary recovery tools. They can be either a long-handled shovel, a military-type entrenching tool that folds for carrying in a backpack or a sturdy pick. Remember, rusted metal can be damaged easily with these tools so use caution when digging your cache so as to not damage its contents.

You'll also learn that hard-packed concrete-like soil is generally an indication that no cache is located beneath it, especially if softer soil is available nearby. Most people are lazy. They would rather dig in softer soil or just bury a cache in a pile of loose rocks.

Most experienced cache hunters go to great lengths to avoid calling attention to themselves. One way to do this is to carry detectors and all other equipment into the field in a backpack. You then appear to be just another hiker. A large backpack will usually accommodate a detector with a 12-inch searchcoil, along with a small shovel and the other tools necessary for an average recovery.

There are numerous reasons for not calling attention to yourself or your search for caches. First of all, you're looking for money; enough said. Plus, you'll be busy and won't need the attention of even honest curiosity-seekers. And, if word ever gets out about your cache recovery, you'll be amazed at the number of people who will try to take it away from you through legal (by claiming rights to all or a part of it) or physical force.

On private property or if there is a question of ownership, negotiate an agreement with the property-owner or individual(s) who might have a legitimate claim to your cache before you begin searching. And, never put your trust in a verbal agreement. A wise man once said that verbal agreements aren't worth the paper they're written on. Also, never leave an open hole after you have discovered something. Even a landowner with whom you have an agreement can get excited about a large hole. He visualizes that it was filled with gold coins that you recovered, and trouble may lie ahead.

When you are working with partners, make certain that all arrangements are made in writing before you start spending money on research and equipment and, certainly, before any cache is discovered. Many of us have had unpleasant experiences, particularly in working with inexperienced treasure hunters. Generally, you can trust an experienced cache hunter who can't afford to have his reputation clouded by a squabble over property rights. Plus, he has handled "found money" before and doesn't tend to get as excited about it.
It's the novice you need be concerned about. Perhaps he simply supplied the tip that began a long and arduous search. Once the prize is recovered, you'll be amazed at how possessive this

person can get about "my" treasure. Why, you may even be offered a "little something" for your time and effort in "helping find it!" Don't ever let this happen to you. Get everything in *writing* before you search.

Of course, even then difficulties can occur. A good friend once told me about discovering a sizable cache on land belonging to a individual with whom he had a notarized agreement. Unfortunately, the landowner saw the cache site before my friend had filled in its hole. Envisioning the hole totally filled with gold, he ignored the written agreement and immediately filed a lawsuit. Fortunately, the written agreement barred the landowner from recovery. Another cache hunter had a carefully worded and signed agreement with a land owner, yet had to relinquish all treasure in the cache because the landowner's wife had not signed the agreement.

Taxes must also be a subject of concern for any successful cache hunter. The Federal Government demands its percentage of income, net, less expenses you derive from treasure hunting just like that from an investment or salary. Similarly, states and municipalities that tax income aren't satisfied until they get their proportionate share.

Yet, who'll know just what you actually recovered out there in the wilderness, much less its eventual worth? That's a good question. Always remember, however, that evading taxes is a crime punishable both by fine and imprisonment. In addition, rewards are given to any individual whose tip leads to the discovery of tax evasion. It's always been my advice, therefore, to pay all taxes that are due and to pay them when they are due. If you can prove that you're in the treasure hunting business, proper expenses can be deducted, but only when treasure is located. Requirements differ from state to state. So, study them carefully to ensure you not pay more than you owe.

Again, my advice to a cache hunter is to keep a low profile in every way. Don't call attention to yourself. Pay your legitimate taxes. Insist on your rights in a quiet, yet firm, manner.

Some of the most pleasant hours I've enjoyed in metal detecting have been spent searching for gold camp caches with my good friends Roy Lagal, Charlie Weaver, Monty Moncrief, Wally Eckhert and Bill Fulleton in the beautiful northwestern United States. Over the years we have hunted for these caches with various kinds of detectors. It's truly amazing how much more effective today's modern instruments are than those we used just a few years back, especially over mineralized soil.

The soil at most of these Rocky Mountain cache sites has a high mineralization content and the terrain is generally rugged. The first challenge for a detector, then, is to achieve precise ground balance that permits faint signals to be heard rather than background chatter. Secondly, searchcoils must be capable of operating at various heights above the ground because of rocks and other obstructions.

Another problem we encountered concerned "hot rocks," those geological anomalies that cause most detectors to give a false signal. Since modern detectors enabled us to deal quite effectively with these little pests, I suggest you employ such a detector in cache hunting, even though you'll be hunting almost exclusively in your All Metal mode. I recommend the Garrett GTI 2500 and the Infinium *LS* models.

Certainly, I don't suggest that when you hunt for a cache, you forget or ignore any of the techniques you have already developed. By all means, use all those special methods that have proved so successful for you and your instrument! As I continue to emphasize in all of my books and articles, basic techniques of metal detecting remain the same because the laws of physics do not change. Rules for ground balancing were valid when you were hunting coins in the park will be just as applicable when you're seeking a cache in the mineralized soil of a deserted Rocky Mountain ghost town. It's the manner in which you apply the basic scanning techniques with a modern detector with a 12-inch or larger searchcoil or a Depth Multiplier that determines whether you can be successful in cache hunting.

Many failures in cache hunting can be attributed to those hobbyists who are thoroughly familiar with the techniques of coin hunting but are inexperienced in seeking the deeper and larger prizes. Because they have full confidence in their detectors to locate deep coins, they may overestimate their abilities to hunt for caches. Since the cache is large, they believe they have all the capabilities needed to locate it. As a result, they envision themselves as cache hunters and conduct proper research to develop a good lead at an old church or mission site used as a hideout after a robbery and the loot, believed to have been buried there, has never been recovered. A great deal of time is obviously required by this research and reaching the site may call for considerably more time and expenses of additional equipment.

Finally, after this expenditure of time, money and emotional energy, the unexperienced cache hunters are on site, ready to scan for the long-lost payroll. Only, the scanning is done with a coin-hunting detector with automatic ground balance in the discriminate mode, using an 8" searchcoil. If the area contains mineralized soil (and this always seems to be true) the instrument must be able to penetrate this mineralization with ground balance equipment capable of compensating for it. The coin hunting detector will leave our unexperienced cache hunters almost helpless...and they won't even know it!. Perhaps they will be able to salvage something from the trip by locating a relic or two.

Occasionally, you can actually see the above scenario portrayed in treasure magazines. The article is about an alleged "cache-hunting" expedition and is accompanied by a photo of the hunters on-site. Look at their searchcoils. If they are small, this hunt may have produced a few relics but probably not a cache...certainly not one that was buried deeply in mineralized soil.

Imagine scanning directly over a valuable cache simply because your detector did not have the proper searchcoil or power or the sensitivity to detect it. Of course, that's exactly what happened to so many of the talented old-timers who used early-day detectors. They didn't even know when they were scanning over the caches that still wait today for our modern 21st-century metal detectors.

Chapter Six
Health, Safety & the Law

The three factors of health, safety and the law should always be considered in hunting for caches. Common sense will usually answer most questions in these areas, but I urge you to always remain aware of them, especially your health and safety and to conduct yourself accordingly.

As for the law, I've always tried to abide by the maxim of that popular old cartoon character, Lil' Abner, who said, "I obeys every law, whether they're good or bad!"

Health

You'll probably be out in the field for extended periods of time. Because cache hunting is not for beginners, however, I'm going to assume that all of you already adhere to these basic rules of healthy treasure hunting:

- Make certain that your equipment is suited to your physical requirements. This particularly concerns the detector's stem, if it is too long you will have a balance problem; it it's too short, and you'll have to stoop over to search.

- Strengthen hand, arm, back and shoulder muscles with regular exercise. Not much is really required here. In fact, just using a detector will probably develop the proper muscles. At the beginning, or after a period of inactivity, however, a hobbyist should protect against strained muscles and ligaments. Remember, the Treasure Hound depth multiplier is heavier than a normal searchcoil. Be sure to compensate for it accordingly.

- Warm-up exercises before each day's activity are generally the answer. Just a few minutes of stretching and other activity to loosen muscles and joints will prepare them for a day's work.

- Finally, during metal detecting activities, use correct scanning techniques and follow accepted rules for stretching, bending and lifting. Whenever you feel yourself tightening up, take a short break.

Most important of all, use common sense and take care of yourself! I know that searching for a big prize such as a cache might tempt you to overexert yourself. Remember, however, that there are no "time limits" to metal detecting. "Your" cache has probably been hidden for a long time, and you probably have adequate time to find it.

Safety

Hunting for a cache, especially in the wilderness, requires more attention to safety than normal treasure hunting with a metal detector.

You must decide for yourself how safe you will be in the areas you plan to search. Accurate knowledge will not only help you dispel many unreasonable fears, but can materially reduce the chances of encountering problems. It is the unknown that we fear most. The best way to avoid trouble is to be ready for it at all times. Remember the Boy Scout motto: Be Prepared.

If you're in the wrong place at the right time, there's always the chance of being bothered by someone. If you lack confidence in the security of an area, don't hunt alone or don't hunt there at all.

No matter where you hunt, always be alert to the possibility of digging up explosives. Over the past half century some areas, especially in the wilderness or "back country", have been used as bombing and artillery ranges. And, in these remote areas you probably won't find posted warning signs. If you dig up a strange-looking device that you suspect might be explosive, notify the authorities immediately. Let them take care of it. Then, exercise caution when digging in that area or just don't search there. The same advice applies to underground cables or pipelines.

Many natural wilderness sites represent a fragile environment that can be easily damaged or destroyed. Even though you may be miles from civilization, please leave only footprints -- not pulltabs, wrappers, cans or other souvenirs of our "disposable" civilization. Remember, a fellow treasure hunter may want to work the area someday. You may even want to come back yourself. Just don't leave your clutter behind!

Once, I was searching for gold nuggets in a Northern Arizona location so remote that I felt certain no one could possibly have worked the area before me. I extended my searchcoil into thick underbrush and heard a loud detector signal. I thought I found a gold nugget because I knew this remote area could certainly not have any metallic trash.

I crawled on my stomach under the shrub and began scraping away loose sand and rocks. At the spot where my detector said, "dig," I recovered a 9-volt battery. And, it was a Garrett battery which my company had ceased to sell long ago! It obviously came from another gold hunter who was not careful about leaving behind his "souvenirs."

When campfires are covered and not doused with water, coals remain very hot even until the next day, which can cause severe burns or start brush fires. Watch out for coals, even when they appear cold.

Toxic waste presents an increasingly serious problem. Be alert to any area (or any piece of flotsam or jetsam on a beach) that looks or smells bad in any way. Keep away from anything that you suspect is contaminated.

Keep track of weather conditions and forecasts to avoid unpleasant surprises. Stay tuned to the NOAA weather radio station in the area where you're searching. You'll need a shortwave receiver tuned to one of these MHz frequencies, 162.40, 162.475 or 162.55. Here, you can listen to continuous weather reporting that includes regular updates and information on unusual weather or storms you can expect. Please don't hunt in adverse weather. Flash flooding and lightning can be hazardous to your health. Radio Shack has good weather radios.

Probably the greatest danger facing anyone in strange places is panic. Such a feeling can easily occur in a deserted area, particularly if you are alone. Sudden overwhelming fear, accompanied by a loss of reasoning contributes to a great many accidents. Try to think calmly about each problem, even before you face it. Condition yourself to resist panic. Remember the single word, "pause." Reacting quickly without thinking usually gains you nothing. Fear can be overcome. Let your reasoning take control to allow you to think your way out of difficult situations.

It's inevitable that people who venture outdoors are going to encounter gnats, mosquitoes, bees, ants, wasps, spiders, ticks, hornets, scorpions and just plain bugs. In the wilderness you might also encounter poisonous snakes and wild animals. You scarcely need to be told to avoid them as much as possible. I've always heard that bears, wolves and the like are more afraid of me than I am of them. It's a theory I've never taken the opportunity to test! I urge you give the right of way to animals and snakes. Now, this talk about insects, snakes and such is not
104

intended to scare you, but it's a good idea to keep the dangers in mind any time you're outdoors. I dug around a large rock which I then rolled from its hole. Beneath it was a giant rattlesnake, but, fortunately, he was too cold to move. While searching Robber's Roost I almost jumped into a nest of baby rattlers. Lucky for me I saw them before I jumped!

Above all, use your common sense, and you'll be fine! Never let needless worry interfere with the joy and thrill of this great hobby. Accurate knowledge will not only help you dispel any unreasonable fears, but materially reduce the chances of encountering problems.

Finally, I repeat the warning I made in *Ghost Town Treasures:* know your companions. Would you believe that people have gone "nuts" at the sight of newfound treasure? I assure you that it has happened! You don't want to be at some far-removed place with an unstable person when a big cache is discovered. I once observed this fellow who became so nervous that his stomach muscles began to jerk visibly and uncontrollably. He perspired profusely and quickly became a basket case. It took him the rest of the night to calm down.

In a similar vein you should always ask yourself if you can really trust the people with whom you may be hunting. I could tell you of several situations when one dishonest person in a group stole some or all of the treasure they were seeking. Select your treasure hunting companions very carefully.

The Law
Once again, cache hunting is different. The everyday hobbyist, looking for coins in the park or on property belonging to an individual who gave permission to hunt, is hardly going to run afoul of any laws that govern hunting for historic artifacts or disturbing potential archaeological sites.

First of all, you must realize that every square inch of property in the United States is *owned* -- by an individual, group, corporation, governmental body, etc. And, there are definitely laws that have been written to apply to various treasure hunting situations. Each state has its own statutes concerning where you can and cannot search and whether you may keep the treasure you find. You must learn these laws, and remember that they can be changed at any time. In searching for big money caches, the stakes are too high to risk losing your prize to a technicality.

Areas such as military sites, national and state parks and monuments are absolutely "off limits" to cache hunting or any other type of metal detecting activity. No ifs, ands or buts here! Not only can your equipment (including automobile) be confiscated but you can face a jail sentence. And, there are other areas where metal detecting is not allowed unless you have permission from local officials. Don't rely on gossip. In fact, if you're really concerned about hunting in an area, it might be a good idea to get your permission in writing from the proper authorities.

All states have laws against trespassing. If a sign says, "Keep Out," do just that. It is always best to seek permission wherever you hunt. Besides, as a good friend likes to say, how can you listen to your metal detector if you have to keep an ear cocked for an approaching screaming property owner, or a siren?

As I stated earlier, laws differ from state to state, but these basic common law concepts generally apply:

- **Finders Keepers** - There may be some truth in this old statement, especially about unmarked items such as coins. But, there are certainly exceptions, particularly when you start considering other objects whose ownership can be more easily identified. No matter what kind of treasure you are looking for, I urge you to have a general knowledge of the laws of ownership. You can never tell what you'll find or where you'll find it! Finders Keepers may not be appropriate for an object you discover on private or posted property if the landowner decides to dispute your claim. On the other hand, Finders Keepers generally applies to any owner-not-identified item you find when you are not trespassing, when you are hunting legally on any public land and when the rightful owner cannot be identified.

Treasure Trove
In the United States this is broadly defined as any gold or silver in coin, plate or bullion and paper currency that has been found concealed in the earth or in a house belonging to another person, even when found in movable property belonging to others, such as a book, bureau, safe or piece of machinery. To be classed as "treasure trove" the item(s) must have been buried or concealed long enough to indicate that the original owner is dead or unknown. All found property can generally be separated into five legal categories:

- **Abandoned Property** - As a general rule, is a tangible asset discarded or abandoned willfully and intentionally by its original owners. An example would be a household item discarded into a receptacle placed at the site of a municipality's normal trash pick-up site such as your home's alley or curbside. If anyone decides to take the item, they can do so legally.

- **Concealed Property** - Is tangible property hidden by its owners to prevent observation, inventory, acquisition or possession by other parties. In most cases, when such property is found, the courts order its return to the original owner.

- **Stolen Property** - Does not necessarily belong to the finder. The person or business from whom the property was taken retains ownership for the life of that individual or institution and then reverts to heirs. Of course, a long stretch of time has passed since many of the robberies discussed in this book. Proving absolute ownership would be a very difficult undertaking.

- **Lost Property** - Is defined as that which the owner has inadvertently and unintentionally lost, yet to which he legally retains title. Still, there is a presumption of abandonment until the owner appears and claims such property, providing that the finder has taken steps to notify the owner of its discovery.

- **Misplaced Property** - Is that which has been intentionally hidden or laid away by its owner who planned to retrieve it at a later date but forgot about the property or where it was hidden. When found, such property is generally treated the same as Concealed Property with attempts required to find its owner. When this is not possible, ownership usually (through the courts) reverts to the occupant or owner of the premises on which the misplaced property was originally found.

Things embedded in the soil generally constitute property other than treasure trove, such as antique bottles or artifacts that might be of historical value. The finder acquires no rights to the object, and possession of such objects belongs to the landowner unless declared otherwise by a court of law.

With the proper attitude and a true explanation of your purpose, you will be surprised at the cooperation you'll receive from most property owners. The majority of them will be curious enough about your metal detector and what you hope to find and agree to let you search. I suggest that you not inform them of the magnitude of your search because in most cases you really won't know this until you actually locate your treasure. Just offer to split, giving them 25% (or less) of all you find and they will usually be even more willing. If both you and the property owner know that a large amount of treasure is being sought, a properly drawn legal agreement is a must! Such an agreement between both you and all landowners (husband and wife, etc.) will eliminate any later disagreements which might otherwise arise. I suggest that you not try to make any agreement with a renter or lessee of property. Locate the owner and make your agreement with him or her. Then, make certain the property owner informs the lessee of your search and recovery plans.

As discussed earlier, the first rule of conduct for any treasure hunter is to fill all holes. You'll learn that most every governmental subdivision, be it city, township, county, or state, enforces some sort of law that prohibits destruction of public or private property. When you dig a hole or cut through grass on private or public property, even in the wilderness, you're in effect violating a law.

In addition, property should always be restored to the condition in which you found it. I have heard of so-called treasure hunters who completely devastate an area, leaving large gaping holes, tearing down structures and uprooting shrubbery and sidewalks. Damage of this kind is one of the reasons we're seeing so many efforts at legislation that would literally shut down metal detectors on public property.

I remember how my good treasure hunting colleague, Ed Morris, was chagrined when he was barred from fair grounds where he made excellent recoveries over the years. His metal detector was banned because other treasure hunters defaced the property. Another friend, George Mroczkowski, was forced to abandon a search because one member of his team dug and left large, gaping holes.

An experienced hobbyist always leaves an area in such a condition that nobody will know that it has ever been searched. In fact, I always urge hobbyists to leave any area they explored in better condition than they found it! All treasure hunters must become aware of their responsibilities to protect the property of others and to keep public property fit for all. Persons who search for valuables by destroying property, leaving holes unfilled or tearing down buildings should not be known as treasure hunters but should be called what they are -- looters and scavengers!

Taxes
It's the law that all treasure you find must be declared as income during the year in which you receive a monetary gain from that treasure. If you find a cache of $1,000 in coins, which you spend at once because they have no numismatic value, then the law requires that you declare the face value of those coins in the current year's income tax report. If, however, you discover a cache of valuable antique coins, you need not make a declaration until you sell the item(s) and then only for the amount you received.

If you decide to donate some of your finds to a historical society, museum or certain other non-profit entities, you may be able to deduct the value (determined by independent appraisal or fair market price) of the items as charitable contributions. Simply stated, the law imposes a tax on all net income from treasure hunting.

In deducting your expenses you must have good records. Check with an attorney or tax accountant, especially if you plan to become a full-time treasure hunter. An accountant will advise you as to what type of records you should keep.

Last, But Not Least

Our hobby, the sport of searching for treasure with a metal detector whether seeking caches in the hills, coins in the park or jewelry on the beach, has been kept clean and dignified by people who care about it, while they express a similar concern for themselves and their fellow man. Most of us who use metal detectors will go out of our way to protect this most rewarding and enjoyable hobby that we love so much, as well as share our enjoyment with others. The simple act of sharing is the primary reason I write books and freely share all the knowledge I have gained over the past 60-plus years. I have left virtually nothing unsaid about treasure hunting because I so greatly enjoy helping others find treasure. As another good friend, famed underwater salvor Bob Marx, likes to say, "There's plenty for all!"

Yet, keeping the hobby clean takes the effort and dedication of everyone, not just cache hunters. So, as you go about enjoying your leisure, or perhaps full-time activity, be professional! Be worthy of this great hobby!

I wish you every success and great happiness. And, I sincerely hope that someday, when we're both out using Garrett detectors to search for caches *I'll see you in the field*!

God Bless!

Chapter Seven
A Mexican Expedition

Nestled deep in Mexico's Sierra Madre Mountains, sits the primitive village of Batopillas. Rich in history this beautiful area of Mexico is brimming with countless treasure stories begging to be proven.

We planned, for some time in the 1960s, a dangerous gold and silver prospecting and mining expedition. There were ten of us including: Javiere Castellanos, Don Garrett (my brother), Curley Jones, Roy Lagal, L.L. "Abe" Lincoln, George Mrockowsky, Frank Mellis, Monty Moncrief and A.M. Van Fossen. Each of these daring men brought with them their countless years of expertise, specializing in metal detecting, prospecting, mining and historical research. Javiere Castellanos, was a Mexican mine owner and prospector who acted as our interpreter and commissioned us to do a metal detector survey of his mining properties.

The centuries-old gold and silver producing areas were about to present the bold group a difficult, yet rewarding journey, as you are about to discover.

Knowing the valley would not give up it's treasure easily was evident as the brave band of anxious searchers slowly and carefully wound their vehicles down the three thousand-foot drop off into the Sierra Madre's Copper Canyon. Primitively paved, the rugged single-lane trail offered no guarantee of safe arrival to their rustic destination below. Reaching LaBufa was a harrowing journey alone without the added stress of having the vehicle's brake lining catch fire. The remainder of the descent resulted in numerous traumatic rides down steep slopes and treacherous hairpin curves.

TOP: This team of professional treasure hunters began their trek through the Sierra Madre Mountain range near Batopillas, Mexico, in search of gold and silver ore during the mid 1970s. Pictured Left to Right: A.M. van Fossen, Roy Lagal, Frank Mellish, Don Garrett, Charles Garrett (sitting), Curlee Jones, Vick Moreland, Monty Moncrief, Javier Castellanos, Arthur. BOTTOM: On the lookout, helping Pancho Villa guard his treasure are modern day gold seekers. Standing L to R: Charles Garrett, Vick Mooreland. Kneeling L to R: George Mrockowsky, Arthur, Frank Mellish, Don Garrett (author's brother) and Roy Lagal.

TOP: What looks like a scene from an Old West film is actually the expedition breaking camp to continue their journey through the Sierra Madres in search of hidden wealth. Though burros were used to carry supplies, much of the journey was made on foot through the steep canyons. BOTTOM: (L to R) A.M. Van Fossen (historian), Mexican miner, L.L. "Abe" Lincoln, Charles Garrett, Javiere Castellanos (mine owner) at the mouth of a silver mine in Southern Mexico. It was during this expedition that Garrett discovered an extremely rich silver vein.

Finally reaching LaBufa offered little solace to the explorers as the valley played host to the notorious vampire bats that have been so vividly featured in publications such as National Geographic. Undaunted by the threat of the frightful bats, the group eagerly explored the area searching out several promising leads. Around the campfire that night, the tired pack began planning their final, 18-mile trek down the Batopillas River embankment on foot.

Beginning the long journey along the Batopillas River's foot and burro path proved just as exciting as the vehicle trek down the mountainside. Several stops along the way provided some long-awaited shady rest and welcome spring water refreshment. The enticement of scaling a large boulder was too much for Van to resist, he began climbing up the massive rock. Looking over the top, he came face to face with an obviously irritated 6' long, open-mouthed hissing iguana that was not about to relinquish his favorite sunning spot. Van peacefully retreated, leaving his new acquaintance to continue his basking session.

Moving down the river, a group of Tarahumara Indians were spotted. They were teaching their children various wilderness survival techniques. One of the Tarahumara Indians offered Van a cache of silver for Van's digging tool.

At that very spot, but across the river was a centuries-old carving of an Aztec Indian chieftain in the process of sacrificing a beautiful maiden by casting her into a deep ceremonial pool. The maiden was wearing a massive amount of native gold and precious stone jewelry. The relief was high on the mountainside, increasing the majesty of the large intricate carving. Later, in a Batopillas home where we were guests, we saw a color painting of this same scene complete with a vast array of treasure about to be sacrificed to the Gods.

TOP: Batopillas, Mexico: The Garrett team spent several weeks in this region searching the Spanish silver mines and other mineral deposits. Narrow roads have since been constructed. In the early days of Batopillas, miners built crude roads for their wagons. One enterprising native, however, bought a WWII Jeep and using a blow torch cut the Jeep into sections. Then he loaded the burros with the sections and transported the vehicle 18 miles down the river to Batopillas. He then reassembled and welded the Jeep together and for many years enjoyed dependable, enviable transportation.

TOP: A once beautiful entrance into this abandoned city, now stands in ruins. It is believed throughout the region that this archway holds the clues to a decades old treasure buried by Pancho Villa. When visiting the site, one can close one's eyes and almost "see" Pancho and his gang of revolutionaries galloping through this ancient entrance way. BOTTOM: Nestled among the Sierra Madres, this abandoned city is a testament of the lives that once lived here. Areas around the city, however, continue to be mined for gold and silver.

TOP: Near LaBufa, Mexico, Charles Garrett examines a cross where several men lost their lives. The robbers stole numerous bags of high graded silver ore and stashed the bags in several locations, which they marked with large painted circles. Upon returning for their silver, the men were killed when their car plunged about five hundred feet down a cliff.

BOTTOM: LaBufa is a quaint Mexican village that rests on the Batopillas River about 350 miles south of El Paso, Texas. Three hundred years ago Spanish settlers discovered the land was rich with silver ore. It was here that Garrett's team of electronic prospectors enjoyed weeks of unspoiled beauty. Mining activities still continue in this region of Mexico. The homes in LaBufa are built on stilts because there is no flat ground. The city is known for a segment National Geographic wrote about its vampire bats, which harass the villagers at night. Treasure hunters know LaBufa because of stories of numerous hidden silver ore caches. The Garrett team parked their vans there because the village, at that time in history, was the the "jumping off place", where only burro or foot traffic was possible during the 18 mile trek to the Batopillas River.

TOP: Javiere Castellanos points to a white circle on a mountain slope. These circles served as clues, or markers, that led to stolen silver caches.

BOTTOM: A.M. van Fossen examines sacks of silver ore nuggets hidden by thieves in the early days of LaBufa. Stolen ore was buried in caves, sealed up and marked with large, round, white circles, which were easily visible in the moonlight when thieves returned to retrieve the ore at night. Van Fossen is a world renowned historian, miner, electronic prospector and treasure hunter.

TOP: Garrett and his expedition made fast friends with many of the native people whom they met during their expedition. Here Charles demonstrates how a metal detector can be used to find gold to an excited group of Tarahumara Indians. BOTTOM: George Mrockowsky, Frank Mellish and Don Garrett watch a helicopter land on this Batopillas River sandbar unloading dozens of soldiers only 100 yards from the Garrett camper van. The treasure hunters enjoyed visiting with the soldiers, swapping stories and teaching them how to use metal detectors as they sat around the campfire each night.

A grueling two days later, the village of Batopillas was finally reached. A beautiful jewel in the heart of this vast, rugged mountain range of silver and gold.

The first place to be explored was a now nameless city that lay in ruins. If time could be turned back two hundred years, this present day lonely gathering of time-worn buildings would be restored to a magnificent city full of wealth and prominence. Though now sadly abandoned of its daily life; this area continues to be mined and produces lavish gold and silver.

So rich was this city, that eighteenth century miners wrote the king of Spain promising him that if he would come to visit their city, his feet would never touch anything but pure silver. This proud claim doesn't even include Pancho Villa's rich caches, gold nuggets, Spanish coins and relics so well hidden, they may never be found.

This desolate city contains many elaborate archways, built to welcome all who entered the grand city. Legend speaks of one of these beautiful archways holding location clues to one of Pancho Villa's valuable treasures. During one of Pancho's many crusades into Mexico's interior, he led his men to this then bustling area. Along the way it seems, he had "acquired" vast treasures, some of which he decided to hide. Since all of the entrances are now unable to be found, the treasure still eludes its seekers. Unexpectedly however, a large underground room at this site was located and explored with ample excitement. In the future, sonar or probing, as difficult as it may be among all the massive boulders, will be necessary to achieve success.

It is believed that this mountain range, or possibly one other, may hold the location of the Aztec ceremonial pool used for elaborate sacrifices to their many gods. In addition, the hundreds of caves in this area could prove to be of great value.

TOP: A joyous Mexican Federale soldier shows his happiness as he waves his rifle above his head, which he had lost during a flash flood. It could not be found, even after the water receded. The soldier was extremely distraught at the time because it is a serious offense for a soldier to lose his weapon. Fortunately, a group of treasure hunters were encamped nearby. BOTTOM: Under George Mrockowsky's direction, they volunteered to do a search and found the weapon buried under several inches of mud. Needless to say, from the happy look on the group's faces, they were all happy for this soldier. George Mrockowsky is kneeling, (front center) wearing a white hat and vest.

On each mountain, in the Batopillas area are several caves perched high upon the slopes. For each one seen, there are probably several more hidden from view. Since the very beginning of time, caves were used as homes and hiding places. This is clearly evident by the painting that depicts a Tarahumara Indian family dwelling. Caves of this kind were used by travelers and outlaws alike to hide themselves or things they didn't want others to find.

Unfortunately, on this expedition the adventurers were mainly interested in treasure cache searches and mine exploration and would have to brave the treacherous journey back some other time to search the caves. Of course now, however, these searches may never be undertaken because the Mexican government controls all mining operations as well as historical, artifact and treasure search expeditions.

Of the numerous caches searched for in the rugged Sierra Madres, the silver ore stash was one that caused a lot of excitement among the anxious band of electronic prospectors. Even though there were several other well-known treasures to be found such as the adobe ghost treasure, the treasure of fire and the Conquistador cave treasure, it was decided to pursue the famed silver ore stash.

Clues were evident along the dangerous slopes of the mountainous region where circular paint symbols had been applied to the rocks as location markers leading to the treasure. As the old story goes, a large quantity of freshly mined silver ore was stolen from a hard rock mine. The crafty thieves buried the silver and marked its secret location with the white paint symbols now found along the roadside. The bandits' plan called for their return a year later to retrieve the stash at night by using low-level light to illuminate the markers. Unfortunately, the criminals celebrated their fortune a bit too early and attempted to retrieve

TOP: A photo taken of the entrance into one of the many mines Garrett and his expedition members explored. The mines were so vast, it was impossible for the explorers to search each one in great detail. BOTTOM: Garrett and a local mine owner with silver ore extracted from mines during their expedition. The silver is 80 to 90 percent pure. Some of this silver ore is on display in the Garrett Museum.

the treasure after becoming drunk. Unable to maneuver the steep slopes and curves at night while in their drunken stupor, they drove their car off a cliff.

There are countless numbers of precious metal mines in the heart of Mexico's mountain range. Mining started in Mexico during the Conquistador era of the 16th century. Both small and large, these mines are an electronic prospector's dream come true. One mine is so vast that a jeep can be driven in a straight line for a distance of one mile through the main tunnel. The biggest advantage in the mining of silver in this day and age is the use of metal detectors. So, with their instruments, the men discovered numerous veins and ore deposits previously overlooked by early-day miners. Ranging in thickness from a sheet of paper to a man's finger at the point of detection, the veins grew many times larger as they snaked deeper into the cavern walls. Any prospector would be elated to be the proud recipient of any one of these silver finds.

In rich mines, its easy to understand why previous miners would discard all but vein rocks that contain pure metal that can be seen by the eye. It appears that as they hacked and blasted their way through the mine following the veins, they did just that. With the use of the metal detectors, the team high-graded old mine dumps and recovered silver that the previous miners simply could not see with their eyes. The powerful detectors penetrated the rock material and could also measure the metal's purity. Sifting through the silver as previous miners had, only the purest pieces were high graded.

Detectors and surface dredges were also used to recover silver ore that for centuries had tumbled down the mountainsides and come to rest in the Batopillas and many other rivers in the area. Due to strict time restraints, only small portions of this silver could be retrieved by the eager hunters.

Deep and multi-stoned, some of these mines run for miles upon miles and must be negotiated with concentration and respect, lest one become lost within their maze. Unfortunately, the group ran out of time and was not able to pursue all the mines this rich area had to offer.

Just one more interesting event to conclude this true tale. Just south of the Batopillas village, the men set up camp near the Batopillas River under large trees. One afternoon the men suddenly heard a loud engine noise. They then saw a helicopter moving toward them just skimming the river's surface. The helicopter landed and out jumped forty Mexican Federales armed to the teeth. Such an event could be nothing but very frightening, but as it turned out, the soldiers were on a month long assignment to ferret out drug runners. They established an encampment less that 100 yards from the Garrett camp. During the following weeks, the prospectors made friends with the soldiers even sharing food and stories and swapping metal detectors for large quantities of silver.

Arriving at the American border on the way home, the Texas Border Guard asked the returning prospectors what they were hauling. The answer, "about a ton of silver ore." Either the guard thought he was being joked with or he wasn't interested in nature's wealth. He replied, "Any liquor?" With a resounding "no", he waved us on our way.

This is but one of the dozens of treasure hunting adventures I made with many of my colleagues over the past 40 years. Virtually every expedition I've participated in, either with family or friends, resulted in improved Garrett equipment designs, features and functions. I've shared virtually everything I know about treasure hunting through countless books I've written over the years.

As I continue to treasure hunt and seek to improve our metal detectors, and the scope of treasure hunting, I hope to *see you in the field.*

Chapter Eight
Outlaws and Their Caches

Now that you know how to research, locate and recover a cache, it's time to consider some of the famous ones allegedly hidden over the years by notorious outlaws. Notice that I used the word allegedly. Anytime you're dealing with a big cache, especially one put down by a famous person, fact and fiction tend to intermingle.

Always remember the Check List I offered on Research (See Chapter Two) before you start spending too much time and/or money looking for hidden treasure. And remember, countless other men and women have likely preceded you in searching for any well-documented cache. Plus, the big prize you've heard about and seek may have been quietly recovered years or decades ago. Who knows? Personally, I believe you'll find it far more rewarding, in money as well as satisfaction, to search for the "little" caches that only you and a few others know about. Even better will be the personal or family cache you seek for yourself by following the directions in the first six chapters of this book.

But, I'm sure the following stories of cached loot will get your adrenaline pumping just as they do mine. Always remember, however, that there are thousands of smaller treasures you can search for while you are on the trail of one of the big ones.

I'm on the trail of about a dozen myself, a $20 million gold coin cache in Idaho, a $7 million gold and silver ore stash in West Texas, a $20,000 gold bar treasure in Mexico, a very small and very old penny, yes, one cent, cache in Garland, Texas, and many, many more in between. If I can do it, so can you!

Yes, treasure hunting is a romantic pastime. And, there's hardly

TOP: A scenic view near Robber's Roost where many early American, Old West outlaws roamed and hid from the law. BOTTOM: A group of electronic prospectors including (L to R) Vaughan Garrett, Dave Loveless, Charles Garrett, Bob Oscarson and Richard Graham search the regions near Robber's Roost. Many of these men are FBI agents who know the area well because they have tracked down modern-day outlaws in this area.

anything more romantic than the thought of finding a vast quantity of outlaw treasure hidden away years ago. So, let's consider some of America's most notorious outlaws and the caches they're supposed to have hidden.

Chapter Nine
Dutch Schultz

Perhaps the most sought-after outlaw treasure of all is the cache allegedly hidden by New York gangster Dutch Shultz in March 1935. As the tale is told a 3x2' metal chest containing gold coins, diamonds, $1,000 bills, Liberty Bonds and other valuables was buried near the village of Phonecia in New York's Catskill Mountains. Depending on who tells the story, the total value of the chest's contents varies from $4 million to $8 million.

This treasure would be almost priceless today because of the rare gold coins alone. The location of this cache, its potential value and the fact that its existence has been accepted for so long combine to earn it the "most sought-after" honor.

You see, literally millions of people live within only a few hundred miles of the site. And, the tale of Schultz burying his treasure has been widely told since he was "rubbed-out" six months after he supposedly hid it.

If a multi-million-dollar cache was reasonably well identified and within easy driving distance, wouldn't you be tempted to give your detector a chance to seek it out? Especially, since you've had so much success finding coins and jewelry in parks and on beaches.

Noted author and social historian E. L. Doctorow, who immortalized Schultz in his novel *Billy Bathgate*, says he doubts the treasure ever existed, but admits that Schultz was "kind of elusive."

Yet, treasure hunters keep flocking to Phoenicia. "The second a stranger comes in and asks for an old map, I know what they're here for," observed Hilary Gold, local librarian.

"Dutch Schultz" was the name the ruthless New York gangster Arthur S. Flegenheimer gave to himself. By whatever name, he was arrested over a dozen times for crimes including assault, homicide, robbery and grand larceny. It was in the lucrative bootlegging racket during the "dry" era of the 1920s that he earned his first big money. After several of his chief competitors were gunned down, he emerged as one of New York's premier suppliers of illegal hooch, both imported and home-brewed.

Even before the end of prohibition closed out the bootlegging business, Dutch expanded into the "policy" rackets of Harlem. It's said that by the end of 1932 he was netting upwards of $20 million per year from various ventures.

Like Al Capone and so many other famed badmen, Dutch somehow neglected to pay taxes on his ill-gotten fortune, which is why the Internal Revenue Service stepped in to charge him with income tax evasion.

Somehow Dutch managed to "beat the rap" in his first trial, but it was obvious that he would someday have to go to jail, if only for a short time, for the crime of income tax evasion.

It was this knowledge of his imminent arrest and incarceration that supposedly prompted the hoodlum to hide a fortune -- safe from rival gangsters and tax collectors -- to await his release from prison.

As the legend unfolds, he and his faithful henchman Bernard "Lulu" Rosenkrantz in late March of 1935 loaded the loot-filled metal chest into their car along with a pick and shovel. They skirted along the Hudson River northward to Kingston where they turned onto Route 28 into the Catskill Mountains. Passing Ashoka Reservoir they drove north, following Esopus creek. At some point south of the village of Phoenicia, they stopped the car and
132

buried the chest near a stand of pine trees near the creek.

It's here the legend becomes interesting. Depending on who is telling the tale, you can hear a lot of versions of just exactly where the cache was hidden. Was it on a straight line between Mount Tobias and Panther Mountain? Was it exactly 4.3 miles south of a specific landmark in Phoenicia? Was it buried buried close to the road because it was too heavy to carry very far? Was it buried deeply because the sandy soil made digging so easy? How about the "map" that Lulu drew and entrusted to Marty Kropier, Schultz's enforcer? Was Jacob "Gurrah" Shapiro responsible for Kropier's murder the same October night Dutch and Lulu were assassinated? Were the deaths connected?

Did Shapiro then recover the fortune? Has anyone since recovered it?

Such questions are just part of the allure of cache-hunting.

Chapter Ten
John Murrell

I'm reasonably confident that I discovered a cache hidden by this legendary bandit who preyed on travelers on the Natchez Trace during the early 19th Century. Unfortunately, I did not take my own advice. I was careless in choosing my treasure hunting companions. As I related earlier, one of them probably returned to the cache site alone. And when we all returned later to recover the treasure, we found only an empty hole.

A fictional account of this treasure hunt is presented in my novel *The Secret of John Murrell's Vault*, with a photograph on the back cover that shows the empty hole.

This cache site near Natchitoches, Louisiana, is but one of the many attributed to Murrell. The son of an itinerant preacher, he was born shortly after 1800 and began his career in crime at an early age, stealing from guests at the roadside inn his mother operated south of Nashville, Tennessee. He soon expanded into highway robbery along the Natchez Trace which ran between Nashville and Natchez on the Mississippi River. This was a busy frontier roadway for travelers and prosperous from goods sold in New Orleans after being barged down the Mississippi. Murrell was usually well-dressed and presented a gentlemanly appearance. He was knowledgeable concerning the Bible and often posed as a country preacher himself, as I recounted in my novel.

In addition to robbery, Murrell also profited from stealing slaves and reselling them. This led to an almost unbelievable scheme he devised for a bogus slave revolt. He formed an organization called the Mystic Clan whose objective was to trick slaves into staging an actual uprising. During the resulting furor he and his gang planned to rob banks, businesses, homes and anything else that held goods of value. After the uprising, he intended to abandon the slaves to their fate and escape with his loot. Murrell's Mystic Clan henchmen stepped up their activities of robbery and slave-stealing to help finance the operation. Of course, no revolt occurred. The plan was thwarted and Murrell was arrested and sent to prison.

While he was carrying out his grandiose project, however, it's believed he cached money and other valuables at various sites used as hideouts by the Mystic Clan. Such locations include: Marked Tree, Arkansas, Plum Point on the Mississippi River near Helena, Arkansas, and in Warren County, Mississippi, on the Yazoo River.

The cache site I discovered and referred to in my novel was made, so the story goes, when Murrell believed that the government was going to prevent the slave revolt by using the Army to break up the Mystic Clan. Believing this, he and his gang transported their plunder to caves near Clarence, Louisiana, on the Red River. He then forced slaves to dig a huge pit nearby north of Natchitoches and actually built a room, lined with hand-hewn stones, to store all this stolen wealth. To avoid detection he then constructed a false graveyard over his vault. The hole I discovered near what looked like a cemetery was indeed lined with hand-hewn stones.

Numerous other cache sites are attributed to Murrell -- near Denmark, Tennessee; Stuart's Island in Chicot County, Arkansas; Woodville, Mississippi; and Louisiana's Honey Island, between the east and west branches of the Pearl River.

I know of discoveries by metal detector hobbyists at two of these sites, but I believe the bulk of Murrell's plundered fortune is still hidden, probably in tunnels and caves or in buried safes. The large quantity of metal in these caches should make them easy to find if a metal detector comes reasonably close to any of them.

As I stated in the Closing Note of my novel about him, I suspect that loot may still be in a sunken vault near the bogus cemetery, and it could be deep.

Did Murrell bury a few small treasures near his underground vault? Did he plan this as a means to fool treasure hunters...satisfy them with just a little wealth? Were the hand-hewn stones I found in the hole not just the platform for a safe but the top of a huge treasure vault? Or, were they the access point to a tunnel that leads to the vault? Remember, he controlled countless slaves who could have dug elaborate vaults and tunnels in which to secret his wealth. Some tunnels have definitely been found not far from the cemetery near the Red River.

Is the treasure still there, awaiting someone?

Chapter Eleven
Sam Bass

Sam Bass was born in Indiana
It was his native home...

So begins the old ballad, which recounts young Sam's dream fulfilled after he journeyed to Texas in 1870.

The story of Sam Bass is popular with me since legend has the outlaw putting down so many caches. Most of them are supposedly located in the Dallas-Fort Worth area. One is told to be hidden on Duck Creek, which marks the eastern boundary of the property on which Garrett's headquarters and factory stands.

Young Sam quickly learned that the life of a cowboy was a hard one and he sought an easier way of making a living. Now, this isn't to say that Young Sam was not a likable and hard-working lad. He merely sought a life that would permit him more time for his chosen pursuits of drinking, gambling and horse racing. With his horse, the legendary Denton mare, Sam was a huge success for a time in the racing business, and he developed quite a reputation as a sportsman.

Of course, the mare's winnings declined as it grew older, and Sam's high style of living demanded lots of cash. Sam began his life as an outlaw after helping drive a herd of cattle north to Deadwood in the Dakotas. When his money ran out there, he organized a gang that tried robbing stagecoaches. Because they found pickings to be slim, Sam decided to begin the career that made him famous, robbing trains. He first chose the Union Pacific Express at the Big Springs station, west of Ogallala, Nebraska. Six bandits stopped the train, on September 18, 1877, but couldn't break open its safe. Instead, they settled for $60,000 in brand new $20 gold pieces from the San Francisco mint.

And, here's the first Sam Bass cache story, the six men divided the loot and split up. They were quickly hunted down and although Sam escaped, three were killed. Concerning the gold coins, it's believed that a good number of them were cached on the banks of the South Platte River near the actual train robbery.

Back in Texas, Sam hid out at Cove Hollow near the community of Rosston in western Denton County, just north of Dallas. From there Sam and his gang launched a 10-month crime spree, and it is at this location that the greatest part of Sam's loot is allegedly hidden.

Again, they first tried robbing stagecoaches, but found the results no more satisfactory than before. Sam turned to trains with a vengeance, robbing them all around Dallas from Allen to the north, Eagle Ford to the west, Mesquite to the east and Hutchins to the south.

Sam and his gang were generally making nuisances of themselves, so sheriffs' posses and the Texas Rangers joined forces to put real heat on them, chasing the outlaws from farm to farm.

Legend tells us that Sam and what was left of his gang finally decided to pull one last big job; they would rob the bank at Round Rock, bring together all their cached loot and flee to Mexico or South America.

The Rangers were waiting for them, however, and as the song says, "*They filled his corpse with rifle balls and emptied out his purse...*"

By this time Sam was something of a Texas folk hero and the

amounts of his legendary outlaw treasure caches grew larger over time. Searchers have hunted over the years at Cove Hollow and at various locations in Dallas such as Chalk Hill, the Trinity River bottoms and Flagpole Hill. He is also reputed to have hidden treasures near Weatherford, Parker County; Bowie, Montague County; Breckenridge, Stephens County; and McNeil, Travis County. Other cache sites are "located" in Wise, Jack, Denton, Cooke and Grayson Counties, all near Dallas. With all of this robbing and burying, Sam must have certainly been an active fellow between the Nebraska train robbery in September and his death the following July!

One tale recounts that before his ill-fated robbery attempt at Round Rock he hid a treasure map showing all the caches in a hollow tree two miles north of town. But, the most famous of Sam's legendary treasures is supposedly buried in Longhorn Caverns, now a famous tourist attraction, which Sam was said to have used as a hideout. Some stories put the fortune here in the millions of dollars.

Where Sam ever got that much money no one seems willing to speculate. Certain so-called Western historians tell us that Sam actually put his hands on very little treasure. Still, you notice that I didn't mention my backyard Duck Creek cache again. I think I'll save that one for myself!

Chapter Twelve
John Dillinger

Who do the people of my generation consider the most notorious outlaw, the most known American criminal of the 20th Century? If answered John Dillinger, no one wold argue with you. His feats of bank robbery, escaping from jail and death stand as a symbol of lawlessness of the 1930s. Yet, he gained this notoriety during a crime spree that lasted less than 15 months.

Dillinger was released from Indiana State Prison on May 10, 1933, after serving more than eight years for a small-time grocery robbery and was later killed in Chicago on July 22, 1934.

During this period Dillinger and his gang stole hundreds of thousands of dollars in cash and jewels in daring bank robberies. These facts are undisputed. The money and jewelry was taken, and most of it was never recovered.

Where did Dillinger hide it? That simple question has intrigued treasure hunters for over half a century. Dillinger's 15-month rampage was so short and so well documented that extensive clues have been readily available to guide searchers for Dillinger's loot.

Three locations in Ohio, Indiana and Wisconsin are traditionally cited as the most likely hiding places.

Chronologically, Ohio comes first. After Dillinger's initial bank robbery in Indianapolis, he was captured by the Indiana State Police. Gang members raided the Lima, Ohio jail, killing a sheriff and freed Dillinger. They hid out at a farm owned by the mother of gang member Harry Piermont and used this farm as a base while robbing ten more banks of hundreds of thousands of dollars.

Known as Hamilton Farm, this cache site is located near Leipsic, Ohio, northwest of Findlay. It is believed that a large quantity of currency and securities is buried in the barn area or in an old cow pasture.

The second cache site is currency buried in the woods surrounding the Little Bohemia Lodge at Star Lake near Mercer, Wisconsin, north of Rhinelander. When the FBI learned the gang took over this lodge, they mobilized a strike force and approached the place. But Dillinger escaped with a small suitcase filled with cash. He reportedly told a confederate later that he buried the bulk of his currency just north of the lodge.

The final cache site is located near Moorseville, Indiana, only a few miles southwest of Indianapolis. On the 10-ace farm of his devout Quaker father, Dillinger is believed to have hidden numerous bundles of currency. The primary cache at this site, however, consists of $300,000 in jewelry taken from safety deposit boxes of a Chicago bank that was going out of business. Dillinger and his gang also took currency and negotiable securities from the boxes and proceeded to his father's farm where he allegedly buried the loot, packed in suitcases.

The Dillinger gang went on a winter vacation to Florida, where police apprehended all except Dillinger who fled to Tucson, Arizona, where he was captured by police and returned to the supposedly escape-proof jail in Crown Point, Indiana. It was here on March 4, 1934, that he made his legendary escape using a wooden gun dyed black with black shoe polish to resemble an Army .45.

Two months later, the Dillinger gang robbed six banks in Indiana and Ohio in only eight days with the final "deposits" made at the Moorseville farm.

In July, John Dillinger's life ended in a Chicago ambush. Or, did it?

Stories persist that the man killed by FBI agents on July 22, 1934, did not even resemble Dillinger. Had plastic surgery after his Crown Point escape changed his appearance? Did the real John Dillinger live to a ripe old age? Who knows.

Stories galore have been told and numerous novels written about his life. We do know, however, that during a 15-month crime spree unparalleled in American history he and a daring gang robbed banks of countless fortunes in currency, securities and jewels. Very little of this was ever recovered. These are facts. The whereabouts of these fortunes is the question that treasure hunters are still trying to answer.

Chapter Thirteen
Joaquin Murietta

Exploits of this legendary bandit of California's Gold Rush days still stir the imagination of treasure hunters and keep them digging in the foothills of the Sierra Nevadas. Joaquin Murietta's reign of terror included robberies and brutal murders and lasted less than two years, but tales of his buried treasure still abound a century and one-half later.

Like that of so many outlaws of the distant past, the story of Joaquin Murietta is surrounded by myth. Even while alive he could not have committed all the crimes for which he and his band were blamed. Nor could he have buried all the treasure attributed to them.

Murietta's history is clouded by folklore and romance. There seems to be no doubt that his real name was Carillo and he came to California from Sonora, Mexico, at the time of the gold rush. One tale says he left Mexico while eloping with a new bride and took a new name to hide from her family as he tried to establish himself as a peace-loving farmer in the Santa Clara Valley. Here he encountered bigotry which resulted in the lynching of his brother and the rape of his bride.

These dreadful events drove Murietta insane, so the story goes, and he vowed revenge over the killers and rapists. He murdered more than a dozen men in this quest for vengeance and -- again, according to one legend -- was driven to absolute madness by the bloodlust he had stirred within himself.

Insane or not, between November 1851 and his death in July 1853, Murietta and his band of outlaws committed crimes ranging from gold hijacking and stagecoach robbery in the Mother Lode country to brutal murder for profit.   The cities of Marysville, Oroville and Paradise first encountered their wrath, but the band of brigands swept across the gold fields robbing and murdering, seemingly at will. Their first stagecoach robbery occurred south of Paradise along the Feather River. Murietta took the strongbox from the San Francisco-bound Concord stage and then cold bloodedly murdered the driver and his four passengers. This Feather River massacre signaled the beginning of Joaquin Murietta's criminal rampage.

As noted above, however, there is reason to believe that Murietta did not commit all the crimes for which he is blamed. During the 20 months of the bandit's terror, it was easy for any robber to declare himself as Joaquin Murietta. Just a few years later, Jesse James saw his reported criminal activities enlarged in a similar manner. In fact, neither the Murietta nor the James gangs could have physically been responsible for all the mayhem and brutality attributed to them.

Still, Murietta and his men robbed numerous stagecoaches and individual miners and murdered victims at will. There is no doubt about that. Whether they took the tens, even, hundreds of thousands of dollars that they are charged with is questionable.
There hardly seems to be a question, however, that Murietta buried some of his loot. Several recoveries have been reported in Calaveras and Shasta Counties over the years, and there have undoubtedly been other finds that went unnoticed.

The caves along the Mercer River near the abandoned ghost town of Bagby are said to contain numerous hidden treasures. Skeletons found in the caves are believed to be Murietta's own men whom he murdered to protect his caches. Many artifacts found in the area, including saddles, gold trinkets, spurs and pistols, have been attributed to Murietta and his gang.
There are a number of areas where Murietta's caches are
144

supposed to have been buried: between Susanville and Freedonyer Pass, near present-day Route 36; between Burney and Hatchet Mountain Pass close to present-day U.S. 299 and in the Downieville area near Route 49. Several caches are suspected between Oroville and Chico near today's U.S. 99 and also along the banks of the Big Butte River between Chico and Paradise. Still another Murietta tale concerns a huge buried treasure in San Diego County near the Mexican border, and treasure hunters over the years have searched around San Luis Rey Mission for hidden loot.

Joaquin Murietta undoubtedly left buried treasure behind him in his bloodstained wake. Recoveries over the years present powerful evidence to support that statement. Has it all been found? Probably not.

Seeking the loot from Murietta's bloody escapades still offers countless opportunities for dedicated cache-hunters. Panning in the rivers and streams, then, is just one of the ways to find gold (or another excuse to visit and enjoy) California's beautiful Sierra Nevadas.

Chapter Fourteen
Jean Lafitte

When all the stories are finally told, I have no doubt that the largest number of outlaw caches, by far, will be attributed to the pirate-patriot, Jean Lafitte. Lafitte sailed from Grande Terre and Galveston Islands in the Gulf of Mexico. When I hear and read about all the places where the pirates of Jean Lafitte and his brother Pierre buried treasure, I'm amazed that they had any time left for pirating, much less helping Andy Jackson whip the British at New Orleans in 1815.

I also remember my friend, the great salvor and diver Bob Marx, telling me that he didn't believe pirates buried much treasure. "They spent it all on gambling, women and liquor," he assured me. Yet, in almost the same breath, he said that he suspected numerous pirate caches could be found in caves on the Caribbean island of Mona. He then suggested that we go there together to find them. Unfortunately, however this island is more likely the resting place of countless unexploded bombs because it was used for bombing practice during World War II.

Needless to say, we've not gone, yet!

Any book about outlaw caches would be greatly lacking without mention of Lafitte and just a few of his probable cache sites.

Lafitte's legend compares to Jesse James' in the number of cache sites he left behind. However, his legend has grown to such proportions that it's hard to determine where the truth about Jean Lafitte leaves off and fiction begins.

Even facts about his life are sketchy. It's believed that he was born in France and appeared in the New World in the early 19th

Century as a New Orleans blacksmith. He also led a band of outlaws based in Barataria Bay offshore Louisiana. His ships flew flags of Central and South American nations in revolt against Spain, and he preyed on both Spanish and neutral vessels.

During the War of 1812 the British offered Lafitte a pardon and a considerable sum if he would aid them in attacking New Orleans. The pirate refused and instead his band of brigands helped Gen. Andrew Jackson's American army win a victory on the plains of Chalmette.

After being pardoned by President James Madison, Lafitte moved his base of operations and returned to the life of a pirate. In 1821 his pirate colony on Galveston Island was destroyed by fire and he sailed away. Most historians believe that he later died in Yucatan, Mexico, or in battle.

But, what about the many caches he is supposed to have left behind?

There's the story of his last plunder of a Spanish frigate from which he took ingots of pure silver. This tale says that the treasure was taken northward by wagon train and sunk in the waters of Hendricks Lake on the Sabine River in East Texas after an attack by the Mexican Army. Various recovery activities have been attempted, and there are even reports that silver bars have been taken from the lake. Perhaps they have.

There's still another yarn about the pirate's "last" treasure cache. After being chased off Galveston Island by the U.S. Navy, his pirate ship, bearing a vast quantity of treasure, was run aground at the mouth of the Lavaca River on the Texas coast. As the ship was sinking Lafitte filled a large chest with gold coins and jewelry, along with sacks of silver bars. The chest was hauled several hundred yards across a salt marsh and buried. Treasure hunters

147

tell stories about how this cache has "almost" been found. Perhaps it has.

Numerous other sites have been searched by treasure hunters over the years seeking Lafitte's buried treasure. Stories recount the recovery of thousands of dollars in gold coins on Amelia Island off the northeast coast of Florida, and a huge cache has been sought near Fowler's Bluff near the Suwanee River in Levy County.

In Louisiana, treasures valued at more than a million dollars are reportedly hidden on Kelso's Island in Cameron Parish and off Malheureux Point in Lake Borgne in St. Bernard Parish.

Other Louisiana treasure sites attributed to Lafitte can be found in Calcasieu, East Baton Rouge, Jefferson, Orleans, St. Charles, St. Landry, Terrebone and Vermilion Parishes.

Lafitte is reputed to have buried treasure in numerous coastal counties in Texas, including Calhoun, Galveston, Jackson, Jefferson, Kleberg and Nueces.

In Mississippi they look for Lafitte's buried loot at a favorite pirate lair on the northern outskirts of Natchez and in Choctaw County.

Did Jean Lafitte leave behind numerous buried treasures? Or, as Bob Marx suspects, did he and his ruffians squander it all in rowdy living? I'd like to believe that treasure was really buried and is still there waiting for our detectors!

Chapter Fifteen
More Pirates

When the average metal detector hobbyist begins thinking of pirate treasure, he visualizes locations in the Caribbean Sea or islands off the coasts of the Yucatan or South America. Most of us think of pirates you see in the context of legendary "Spanish Main" buccaneer vessels raiding Spanish galleons as they left Mexico with Montezums'a treasures.

Or, we remember Jean Lafitte who's discussed in the previous chapter of this book. Treasure caches attributed to him are hidden on beaches from Florida to South Texas.

But there were other pirates plundering vessels on the high seas in America's early days. Waters off the East Coast were the hunting grounds of many successful and notorious pirates. These ocean-going thieves are believed to have buried their treasures far from Gulf of Mexico beaches and islands in the Caribbean. During the early part of the 18th century literally hundreds of pirates operated along the Atlantic coast. Captain Kidd and Blackbeard (Edward Teach) and Black Sam Bellamy are three of the most famous.

Captain Kidd

For more than two centuries Capt. William Kidd has been the storybook pirate. With a commission from King William III to privateer against the French, he was charged in 1669 to "rid the seas of pirates." Instead, he became a pirate himself. The why and wherefore of such a transformation, or, if it even occurred, have been told time and again.

What is known is that Kidd rather publicly buried a small treasure on Gardiner's Island in Long Island Sound shortly before he was arrested to be tried in England for piracy. This treasure was recovered, but most historians regard it as a ruse designed to conceal the burying of larger treasures.

Where were they buried and do any of them still exist? Block Island and coves of Maine and islands off its coast and that of neighboring New Hampshire are cited as possible caches. A popular story holds that the famed Astor family gained its first wealth by digging up Capt. Kidd's treasure on Maine's Deer Island. Treasure hunters also sought Capt. Kidd's plunder on New Jersey's Absecon Island, in Delaware at Cape Henlopen and Kelly Island and near Edgecombe, Maine.

It is known that before he was hanged, Capt. Kidd tried to bargain for his freedom by relating the location of various caches.

Also, while in jail in Boston he smuggled a letter to his wife which allegedly told her -- in code -- how to recover his treasure. Edgar Allan Poe devoted his book, *The Gold Bug,* to his efforts to decipher this code.

Capt. Kidd certainly left behind a legend that few other outlaws can match. Whether he left behind buried plunder as well is a question that still tantalizes treasure hunters.

Blackbeard
The caches of Blackbeard have been located along the Atlantic Coast from Georgia's Ossabaw Island to the coves and islands of Maine. Mulberry Island (probably known by another name today) has been noted as the Maryland site of a Blackbeard cache.

Black Sam Bellamy gained notoriety in recent years when his sunken pirate ship Whidah, sunk in a violent storm in 1717 while loaded with plunder, was discovered by Barry Clifford with the loot still aboard.

A popular location for Black Sam's largest cache is the mouth of the Machias River on the southeast coast of Maine.

Dixie Bull

Dixie Bull is considered by many to be New England's first pirate in the early 17th century. Most popular locations for his treasure caches are on Damariscove Island just south of Boothbay, Maine, and Cushing Island, just off Portland, Maine.

Billy Bowlegs Rogers

Billy Bowlegs Rogers began his career as a pirate sailing with Jean Lafitte pillaging the Gulf Coast of Florida in the 19th century. However, unlike other pirates he lived to a ripe old age. Tradition maintains that Rogers either buried or lost loot at six different locations along the Florida coast, and silver bars have been recovered at Fort Walton Beach. Of course, these legends put the value of Billy Bowlegs caches in the millions of dollars.

Florida locations given for these caches are somewhere in Franklin County, Santa Rosa in Walton County, Fort San Carlos near Pensacola and Bald Point in Apalachee Bay. His last ship, allegedly filled with coins and other treasures, sank in shallow water at the mouth of the Suwanee River and for some reason was never salvaged by the pirate.

Most historians question some of these "facts." Billy Bowlegs lived to the age of 93, dying in 1888, and did not live in luxury. We are told that he spent most of his life in a cabin at the mouth of the Suwanee keeping watch over his sunken schooner. Why didn't he dig up the treasures himself and enjoy 19th century luxuries? Did the buried treasures even exist? Are any of them still at the traditional locations? This last question is one that only a dedicated cache hunter can answer.

There are many, many other pirates whose stories make good

reading and whose legendary exploits are said to have left behind buried treasures galore. Some of them were keelboat pirates who buried their treasures on the banks of the Mississippi and Ohio Rivers. Well into the 1880s the San Juan Islands in Washington's Puget Sound served as bases for numerous gangs of pirates who -- you guessed it -- left behind treasure caches.

Black Caesar

In Florida, Henri Caesar, known as Black Caesar and Caesar the Great is reputed to have abandoned a multi-million-dollar cache of gold, silver and jewelry on Sanibel Island and another treasure at Black Caesar's Rock in Dade County.

The list of pirates and their "buried treasures" seems to be endless. I continually hear of someone new who plundered on the high seas centuries ago and left his treasures behind for you and me to find. I hope we can!

Chapter Sixteen
Pancho Villa

Strictly speaking, the great Mexican revolutionary general, Pancho Villa, was not an outlaw who would have buried his loot. But some of the greatest cache stories of our era are linked to this legendary leader. Was he a robber-bandit? He claimed not to be, saying he only took from those who had much and gave to those who had little. Plus, he said what he seized from banks, mines, estates and the like only belonged to the people and was needed to finance their revolution against dictatorship.

A number of years ago the El Paso, Texas, Chamber of Commerce gained national publicity with a complicated tale of Villa's buried treasure in the nearby Franklin Mountains. The public was invited to come search for it. I guess they were also welcome to spend their money with El Paso merchants. And, there are scores of other tales concerning buried treasure left behind by this legendary figure.

Of course, it's been my experience that every adult (and, many children) in Mexico has a cache story to tell you. He or she knows about a treasure that required hiding during some period of Mexico's turbulent history.

What makes my Pancho Villa cache story so different from the others is that I heard it from the general's widow herself. She spoke with absolute faith in the existence of hidden treasure at several locations. Plus, she let me use a metal detector to search the general's house in Chihuahua City.

That's right, Garrett metal detectors found indications of treasure in Pancho Villa's home. It was in the early 1970's when I went with a group of treasure-hunting friends on a month-long trip to the

ABOVE: A painting of Pancho Villa, the Mexican revolutionary general who is known to have cached vast amounts of treasure throughout Mexico and parts of Texas.

TOP: The Garrett van is parked in front of Pancho Villa's estate (also used as a museum) in Chihuahua, Mexico, where Garrett and his expedition met Señora Villa, who gave the group a tour of her home. BOTTOM: Inside the Pancho Villa estate. Señora Viilla permitted Garrett to explore the museum in search of caches her husband may have hidden in the home. Note her portrait to the right of the door.

TOP: The courtyard at Pancho Villa's estate where Garrett believes a treasure may have been hidden by Villa himself. See BOTTOM photo on Page 113. BOTTOM: The car in which Pancho Villa was riding when he was ambushed. It has dozens of bullet holes.

156

interior of Mexico. We planned to search for treasure and scan in numerous old Spanish silver mines...all of this in mountain wilderness hundreds of miles south of the border. We were to find silver nuggets the size of a man's fist and ore veins of almost pure silver a foot in width as well as a few caches. We even examined a ghost's haunt that can still cause the hair on the back of my neck to stand up like porcupine quills. And, if that's not enough, we learned from eyewitness accounts of a strange Mexican fire that surely marks the location of buried treasure.

Before going on to our location near Batopillas, however, we stopped in Chihuahua City to stock up on food and other equipment. While we were waiting, our Mexican associate, Javier Castellanos, asked if we would like to meet his friend, the widow of Pancho Villa, the lovely Señora Luz Corral Viuda de Villa. He told us that the home in which she lived also served as a museum.

We were met at the door of a handsome two-story stucco structure designed in the sprawling open style popular in Northern Mexico and the Southwestern United States by a Mexican woman who appeared keen and alert, though clearly elderly. She dressed elegantly and presented a stately appearance. Sra. Villa was a lovely lady and it was obvious that she had once been a very beautiful young woman.

Sra. Villa showed keen interest when she learned details of our proposed search for ore and other treasures. She was quite familiar with our Batopillas destination since her husband had campaigned in the area. She paid particular attention to the description of metal detection equipment and asked for a demonstration.

Sra. Villa observed with apparent delight as we explained metal target differentiation and used Garrett detectors to show her how

157

ABOVE: In the Pancho Villa home and museum in Chihuahua, Mexico, the Garrett team visited with Ms. Pancho Villa. In this photo members of the treasure hunting team demonstrate to several people the capabilities of detecting equipment and how the instruments can differentiate between coins and iron items. Ms. Villa had been watching the men when suddenly she walked over and joined the group. She watched for a while and then removed her gold pendant and chain, then laid it on the floor, Javierre, who had been demonstrating the detector, scanned the searchcoil over the pendant. Each time he did so, the detector sounded out with a loud signal. A smile came over Ms. Villa's face. She then motioned for her grandson, who was standing with her, to pick up her necklace. Ms. Villa asked Charles to travel with her to search for two of Pancho's treasures.

discrimination permitted detection only of those targets made of precious metal.

Sure enough, we then began talking about treasure hunting and "swapping lies" as those of us in the fraternity are prone to do. Finally, I asked her if she believed any of the stories about her husband's caches could be true.

"About the stories...quien sabe? Who knows? About the money caches...yes, I know where some are located."

"You believe you know where he hid treasure?" we asked her. "Is that what you're telling us?"

"No, I do not just believe. I know the places where my husband hid treasure. Now, among the things that I believe is that he concealed treasure elsewhere...even within the walls of this house. That, I do not know for sure."

When she said this I immediately asked if she would permit us to search here...today...with our metal detectors.

Javier was startled and began apologizing for me, but the gracious señora said, "I told you that mi casa es su casa, and I meant it. It would be my pleasure to watch you search for any treasure that my husband has hidden here."

We selected our best detectors and I used one with the ability to penetrate all types of material while ignoring most nails and construction trash as well as the metal screening that can always be expected in stucco construction.

We began in the museum's front room which had a wooden floor and walls of wood and stucco. We scanned every room in the museum and living quarters, keeping records of all detector signals. When the search continued in the rear and remote parts of the old building, we encountered something under the tile floor that sounded suspicious. Contouring enabled us to determine that the target was between 12 and 18 inches in diameter and was buried at least two feet deep.You can be sure that we wanted to put a sledgehammer to the tiles.

We continued our search along various walls until I reached the entranceway to an enclosed patio of the dwelling where my Garrett again rang out clearly and positively. It was a good signal, and we scanned the area carefully, trying to determine both the size of the target and its depth in the wall. Making this location even more exciting were indications here of replastering after the wall had been originally built. We finally determined that it was about quart-sized and just a few inches deep in the wall.

We presented our findings to Sra. Villa in a large courtyard with a big ornate fountain several yards in circumference. Looking down on the fountain from a height of about 8 feet was a large bust of the revolutionary general on a pedestal.

Unfortunately, it was already past the hour when we determined we must leave. Although we took only a few moments to inspect the courtyard with detectors, I would have liked to have spent more time scanning around the decorative fountain overlooked by the bust of Villa. It was obvious that it would be brightly lighted at all times, so I strongly suspect that the treasure is buried beneath the pedestal upon which Pancho sits in all his glory. The other two treasure sites in the house were merely diversions.

Every time I think of this I laugh as Pancho Villa himself must. Just imagine. There he rests to this day, his likeness molded into a smiling bust, sitting atop a king's throne pillar with an eternal smile as he guards a fabulous treasure which he will open in his eternal afterlife.

But, there was no more searching for us that day. We had too many chores necessary in preparing for the trip to Batopillas. Our hostess was obviously interested in the location of the detector signals and the estimates of target size. "What a shame that you must leave so soon," she declared. "Why, you certainly have my permission to open the floor and wall where you think treasure might be hidden. I would be happy to share anything you find. When you return after your adventure at the mines, I would also like to give you information about two other treasures hidden by Pancho Villa."

We stayed at the mines longer than we had planned and really had no time even to pause in Chihuahua City on our way home. Still, we had promised Sra. Villa. So we took time to visit and delighted her with several gifts, including pieces of pure native silver that we had discovered with our detectors.

When she realized that we would not be able to search her home further, she called me aside. Because she said our discussion was quite important, she asked that our Mexican friend interpret, in case her English faltered. Her language was plain and clear. In fact, what she told me that day still sends chills up my spine!

She asked me to accompany her to two sites, and she asked me to bring along my best metal detector. One location was in San Antonio. The other she would identify only as "the interior of Mexico."

When I asked if the Mexican cache could be found in the Durango mountains or the Chihuahua desert, she offered only a Mona Lisa smile and murmured, "So, you are familiar with those legends. Well, one of them is true."

She was insistent in exacting a promise from me to return soon for the two journeys...one that would take us to San Antonio where she gave me the site of a buried treasure and the other to a location she continued to describe only as "the interior of Mexico."
A treasure hunter's dream come true...to go hunting for the caches of a bandit...with his wife as guide!

As we left Chihuahua City, I could think only of the adventures that awaited my return to Sra. Villa...a reunion that, alas, was never to take place...destiny and death shortly overtook the lovely señora.

Such a misfortune, but not the first! I believe that I would need the fingers of both hands to count the times I have postponed a treasure search for too long. Take my advice, don't delay when you have an opportunity to follow an important lead. You may never get another chance!

Chapter Seventeen
The Ashley Gang

Except for that inquisitive breed known as treasure hunters, few in Florida today are familiar with the exploits of the outlaw band led by John Ashley. For more than a decade, during the early 20th Century, this family of Florida "cracker" brigands ravaged their way across the southeast corner of the state, laughing at jails, warrants and posses.

The Ashleys' began their criminal activities as petty thieves and hold-up men, but soon graduated to bank robbery and murder. The coming of prohibition offered new opportunities and the renegades became notorious for hijacking bootleggers while occasionally engaging in rum-running themselves. Their success as criminals was due, in no small part, to their intimate knowledge of the Everglades, that vast trackless swamp as impenetrable as any forest in the world. Lawmen were unable to follow them after they retreated to their hide-outs there.

Operating first from a country store and filling station near the village of Canal Point on Lake Okeechobee's south shore, the gang was led by John Ashley. Others included his brothers Bob and Bill, sister Mary and her husband Hanford Mobley as well as John's "moll" Laura, self-styled "Queen of the Everglades."

Of course, legend has overtaken the facts concerning exploits of the Ashleys, but it is recorded that they robbed banks in Stuart, Pompano and West Palm Beach and numerous other crimes.

Their prohibition-era exploits of hijacking and piracy have similarly been upgraded to legendary status so that it is difficult to separate truth from myth. What is known is that they seized numerous loads of contraband across the southern part of the state and on the high seas at gun-point, while they brought in illegal liquor themselves by boat from the Bahamas.

The result of all of these crimes naturally was money, far more than they ever could have spent. As notorious outlaws they were forced to remain in hiding essentially throughout their decade-long career. Where they hid this loot is the question that has captivated the imagination of treasure hunters for more than seven decades.

Two shoot-outs with lawmen in 1924 ended the Ashley era in Florida's annals of crime. The first came in January when lawmen ambushed gang members north of West Palm Beach, killed John's father Joe and captured seven other members of the band. One tale is told of a gang member then offering to lead lawmen to the hiding place of the Ashley treasure in exchange for his freedom. The authorities agreed, but instead of directing officers to the cache, the bandit led them to an isolated area where his henchmen were waiting to help him escape.

It was in November that the final chapter was written on a bridge over the Sebastian River. A car bearing Joe Ashley and three henchmen was stopped at a roadblock, where all four were killed in gun battle. Folklore insists that they surrendered and were then handcuffed and shot by the lawmen. How they died matters little to us treasure hunters. We are concerned with all the fortunes they stole and never spent.

Two sites are usually given as the resting place for the Ashley loot. One is near Canal Point on the banks of Lake Okeechobee where John and Laura lived for a time. A more popular location is an island in the St. Lucie inlet, with the site located just west of the mouth of the inlet or a few miles south on the Intracoastal Waterway near Peck's Lake.

Vast areas of the Florida Everglades remain today as they were when the Ashleys roamed them in the 1920s. They have changed but little. Perhaps it is here deep in the swamp that an outlaw treasure is awaiting discovery.

Chapter Eighteen
Other Outlaws

One of the great pleasures of writing this book has been the opportunity to review some of the interesting cache stories about America's most notorious outlaws. And, believe me, it seems that each one of them is said to have left behind a cache somewhere.

Previous chapters have featured well known outlaws with either a single documented significant cache or numerous hordes of buried loot. I'd like to review for you now the caches attributed to a few other infamous desperados and their gangs.

The Dalton Gang

The infamous Dalton Gang, whose exploits ended when four of them were killed in the famous Coffeyville, Kansas, gun battle in October, 1892, left behind only a single cache story with multiple endings. Of course, the question even today remains exactly what happened to the loot stolen in various bank robberies before their fiasco at Coffeyville.

One version has them riding from Tulsa to Coffeyville and burying some $10,000 at one or more of their campsites along the way. These camps have been variously located near Nowata at Onion Creek, Lightning Creek and California Creek.

One Dalton brother, Emmett, survived the gun battle and served time in prison. Some folklorists believe, however, that he never tried to retrieve any of the caches. Still another version of the Daltons' cache says that Emmett was not with the gang when they buried their loot on a hill north of Pleasonton, Kansas, near the old road between Fort Scott and Trading Post. This yarn says that Emmett made numerous efforts to locate the cache but was unsuccessful.

Al Spencer

This same Oklahoma-Kansas area is the location of another tale of missing outlaw loot, that of Al Spencer. He is said to have "robbed more Oklahoma banks than any other bandit." The cocky little "King of the Osage Badlands" indeed robbed numerous banks and at least one train in the early 1920s. And, the fact remains that when a posse killed him he was carrying $10,000 in bearer bonds, only half of the loot from a recent train robbery. There's the rest of Spencer's ill-gotten gains? Legend tells us that

ABOVE: Nighttime approaches as Roy Lagal, Tommy T. Long and Charles Garrett began their detector search of an area where several outlaws hid a stash of gold in South Dakota. It was not known how deep the cache was buried but the searchers know the Depth Multiplier offered the fastest scanning speed and they were not interested in anything small.

Twenty minutes into their search their detector signaled for the men to dig. A few shovels of dirt later a large rock was exposed. Shortly after they had completely uncovered a five-gallon sized rock and began rolling it out of the hole. Tommy T suddenly yelled, "Get back, get back, snake, there's a snake!" In less than no time the men cleared the snake's home almost falling over the large rocks that lay nearby. Cautiously they eased forward and peered into the ground. A large diamond back rattlesnake lay coiled apparently enjoying its winter slumber. Further cautious and careful scanning and digging revealed a very old army canteen.

ABOVE: This safe, drilled and pried open, was found in a small North Texas pond. Charles Garrett sponsored an underwater metal detector seminar for Garland Police. To practice with underwater detectors, the men scanned the lake's muddy bottom. They found several bicycles, a motorcycle, several vending machine coin changers and this safe. The recovered items were all stolen. The safe contained a large amount of cash and other items.

168

it is still hidden somewhere in the rugged Osage Hills west of the Verdigris River in northeastern Oklahoma.

Al Capone

How about famed "Scarface" Al Capone, reputed to be America's richest gangster? He can't be left out of a book about bad hombres. Treasure hunters who have tried to trace his activities believe that he buried one or more large caches near his secluded Palm Island estate in Florida, near Paul Beach. It was here that he died in 1947. They believe he put the money down before going to prison for tax evasion in 1930. And, like all the other cache stories in this book, it's still there! It's my guess that he hid his wealth in some structure of his Florida estate. I don't believe he would have trusted the shifting sands.

Bonnie & Clyde

Bonnie (Parker) and Clyde (Barrow) were notorious even before the movie gave them worldwide fame some years back. There's a cache story about them too. It seems that they, along with Clyde's brother Marvin, his wife and their teenage car thief pal were camping in a wooded area overlooking the Raccoon River about three miles north of Dexter, Iowa. An ambush by lawmen occurred here (not in a tourist court, as shown in the movie) with Clyde's brother killed and his widow wounded and captured. The wife, Blanche, told authorities that the gang hid their plunder amounting to several thousand dollars in the woods near their campsite. She went to prison, Bonnie and Clyde died in a Louisiana ambush and, once again, there's never been any report of their cache being recovered.

Henry Plummer

Henry Plummer robbed his way through the Old West in the 1860s and died at the end of a vigilante rope. During the gold rush days of Virginia City, Montana, he was even elected sheriff! But that didn't stop him and his gang from robbing miners, gamblers and other unfortunates who mined a little gold. Finally, townspeople became irate enough to put a noose around Plummer's neck.

As he was about to die the robber/sheriff is supposed to have pleaded with the mob to free him so that he could lead them to a cache of $300,000. Four ghost towns are variously given as the possible location of Plummer caches. They are Virginia City, Bannack, Deer Lodge and Haugan.

A postscript to the Plummer story tells of a woman with a treasure map who in 1888 reportedly recovered four sacks of gold buried by Plummer. While she and her sons were crossing the Dearborn River south of Augusta, a highway man shot at them and killed the horse packing the gold. The highwayman fled, and the woman and son found the dead horse. But the gold was missing, and legend has it that it still is!

Fred Burke
Fred "Killer" Burke of St. Valentine's Day Massacre fame was called by police of the early 1930s "the most dangerous man in America." His crimes ranging from kidnaping and bank robbery to various holdups netted him and his gang hundreds of thousands of dollars.

With this accumulated wealth Burke decided to go into hiding. He feared not only lawmen but rival gangsters. He traveled to northern Missouri and found work on the farm of Barney Porter, located northeast of Milan, about halfway between that town and Green City in Sullivan County.

Burke's life outside of crime was short-lived. He was surprised by lawmen while asleep and rushed to St. Joseph, Michigan, where he was tried for the murder of a policeman. Burke was sentenced to the state's maximum term of life imprisonment, and you know the rest of the story. His loot was never recovered and may still lie buried in various caches at the site of old Porter Farm.

Roger Touhy

Tough Roger Touhy reportedly buried more than $50,000 near Rockford, Illinois, and another $60,000 was allegedly cached by his gang in the Cosby area of Newport, Tennessee, east of Knoxville. Touhy was gunned down by rival mobsters before he could recover any of this loot.

Butch Cassidy & The Sundance Kid

Did Butch Cassidy and his Wild Bunch pals leave behind any caches? A man who called himself the son of Harry Longbaugh, the Sundance Kid, claimed that they did and offered tattered old treasure maps as proof. He alleged that nearly half a million dollars was buried in 21 caches in Montana and 14 in Wyoming.

It's known that the Sundance Kid always exchanged paper money for gold and silver, and he's reported to have made several mysterious trips from the old log Hammond homestead located where Sweetwater County, Wyoming, adjoins Moffat County, Colorado. Gang members believed that Sundance had a cache somewhere within four miles of the homestead and Powder Springs.

The Great Northern train robbery at Wagner, Montana was the last crime committed by Butch and Sundance before they left for South America. It's reported, however, that several members of the gang traveled to Idaho "with six or seven sacks filled with gold." Some believe that this gold is still cached along a creek east of Spokane, Washington, in Kootenia County, Idaho. Cassidy robbed a payroll in Carbon County, Utah, and fled to the bleak San Rafael swell where he is supposed to have buried the loot in Buckhorn Canyon after having his horse shot from under him.

There are stories of other Wild Bunch caches in the badlands, blind canyons, hidden parks and wind swept mesas of the San Rafael swell and Robbers Roost (Southeast of Utah) areas which I visited and researched when writing *Ghost Town Treasures*. These are certainly desolate regions void of any semblance of civilization.

In the mid-90s when I made a trip to Utah to test new equipment under field conditions and conduct research for my book Ghost Town Treasures, we spent several days exploring the famed Robber's Roost in the southeastern corner of the state. The Roost area includes deep, eroded canyons and fortress-like cliffs that did indeed give bandits a good hiding place with a "roost" to inspect the approach of any lawmen or others. This strangely beautiful and isolated high desert country was one of the last fully explored areas of contiguous lower 48 states; Scenic beauty truly defines the region.

Left: Garrett searches near Robber's Roost

Opposite Page
TOP: Garrett searches an abandoned site with a hand scanner

CENTER: a scenic view of the area where many outlaws hid from the law

BOTTOM: Garrett searches caves where outlaws once hid

The area of the Dirty Devil River and north of the Colorado was known as Robber's Roost long before teen-age Butch Cassidy drove his stolen horses three in 1884. Cap Brown, a notorious horse thief of the 1870s was one of the first outlaws to find sanctuary there.

I can certainly attest that the desolate regions south of the bleak San Rafael swell is rugged indeed and pretty much devoid of any semblance of habitation. There are numerous caves, rocky crevasses and other formations where outlaws could hide.

Did butch Cassidy and his wild bunch leave behind any caches in Robber's Roost? It's known that the Sundance Kid always exchanged paper money for gold and silver, but these members of the Wild Bunch were also known to squander money recklessly every time they completed a job. Perhaps there was nothing left for them to cache.

Cassidy robbed a payroll in Carbon County, Utah, and fled to the arid San Rafael swell where he is supposed to have buried the loot in Buckhorn Canyon after having his horse shot from under him. There are stories of other Wild Bunch caches in the badlands, blind canyons, hidden parks and windswept mesas of the Swell and Robber's Roost.

You can read more about this in Ghost Town Treasures. And see some of the country for yourself in our video, Utah Trek, filmed during the trip. Our own "Utah Bunch" that participated in the trek included Dave Loveless, Richard and Norma Graham, Stan Bell, Bob Oscarson, Sheriff Mike Lacy (well known throughout the Four Corners area) and others who we especially thanked for their patience and graciousness in hosting the event.

Billy the Kid

Billy the Kid, famed killer New Mexico's Lincoln County war, left us only one cache story. It's worth considering simply because of the notoriety associated with him. The cache was "only" a single Colt revolver that he had used in his Lincoln jailbreak, but think what it would be worth today!

Billy was in the Lincoln County jail waiting to be hanged after his conviction for murder of a sheriff and his deputy. On April 28, 1881, just two weeks before his date with the gallows, Billy killed two sheriff's deputies and escaped. He stole a horse and fled Lincoln with leg chains dangling from each ankle. He headed west toward the Capital Mountains where he knew that friends would help him. After passing through Baca Canyon, Billy abandoned his stolen horse for some reason and proceeded on foot. Chains on each ankle slowed him down, and his arsenal of two six-shooters and .44-cal. rifle plus ammunition was heavy. While hiking through Capitan Gap Billy decided to lighten his load and carefully hid one pistol in the fork of a juniper tree just off the trail north of the divide's crest.

When he reached the Las Tablas community Billy suggested to a friend there that he retrace the path he had taken, retrieve the pistol and keep it. Which is exactly what that friend tried to do only, he couldn't find the gun, even though he searched for it several times. There were too many junipers in too many gullies just like Billy's description.

When the young outlaw was finally killed (by Pat Garrett at Fort Sumner, of course) on July 14, the friend, plus his friends, became even more determined to find Billy's pistol. They recognized its value. Still, they had no luck. And, neither has anyone since. At least, if so, they're not talking about it. And, I think they would certainly boast of this valuable relic or sell it.

Ma Baker

One last cache story concerns the famed gang of Ma Baker and Alvin Karpis. Against Ma's wishes the gang kidnaped St. Paul banker Edward G. Bremer in 1934 and received ransom money in excess of $200,000 in small bills. The ransom was paid and Bremer was released in Rochester, Minnesota. About half the money was later recovered or accounted for, but legend has it that the other half was buried in a strongbox by Fred Baker along the road leading south of Rochester to Chatfield.

Fred and his mother were killed a year later in a machine gun battle with federal agents in Florida. Fred was carrying $4,000 of the ransom money when he was killed, but there has never been any report of the missing $100,000. Perhaps it's buried along that Minnesota road in a metal box that's just waiting to give signals from a metal detector.

These stories about outlaw caches could go on endlessly. Almost every desperado of any repute is alleged to have "buried and never recovered" some part of his ill-gotten gains. And, you know what? There's bound to be some truth to the tales!

Chapter Nineteen
Jesse James

It's appropriate that this book end with the story of Jesse Woodson James, one of the most famous desperados of them all. He was truly a legend in his own time. And, this legend which has lived for more than 150 years presents him as the all-time champion in putting down caches.

Jesse is another outlaw whose facts and myths have touched my own life and that of Eleanor, my wife and treasure hunting companion of 50 years. One of the tales concerning Jesse is that Bob Ford, the "*dirty little coward who shot Mr. Howard*" did not put poor Jesse in his grave as the song relates. A man who lived to a ripe old age and is now buried in Granbury, Texas, just southwest of Dallas, claimed to his death that he was Jesse James and that the shooting in 1882 had been arranged to stop lawmen from pursuing him.

Oh, I know, that the remains of "Mr. Howard" were exhumed in 1997, and DNA testing proved them indeed to be those of Jesse Woodson James. Let me quickly state that I don't deny the effectiveness of DNA testing.

But, I'll confess, I love a good tale. And I know my father , Wayne Garrett, introduced me to an elderly friend of his, Cole James, who was named after Cole Younger, a believed outlaw and friend of Jesse James. Cole claimed to know the real Jesse James (whomever he was) more than half a century earlier. While my father and I were visiting with Cole James at his backwoods home deep in the East Texas piney woods, in 1973, Cole described how Jesse and his companions stayed in nearby woods until they saw a white sheet hanging on the front porch. This sheet was the signal that it was safe to come up to the house.

TOP: Wayne Garrett, the author's father, discusses the life of Jesse James with Cole James on the front porch of Cole's East Texas home. BOTTOM: Wayne Garrett points to the wooded area where Jesse James and his gang hid before approaching the house.

Cole James also told us he knew that the elderly man in Granbury was actually the real Jesse because of scars on his arm. He claimed that during a visit in the early 1900s when Jesse showed the scars to a James relative, she exclaimed, "My God, you're Jesse James." Cole told us she recognized the scars from a knife wound that she had sewn up with needle and thread many decades ago.

This friend of my father swore that the man buried in Granbury was the "real" Jesse. And, members of Eleanor's family (her father, Mervin Smith and Uncle Elmer Smith, both talked with me at length about Jesse's life in East Texas) relate similar experiences. Jesse James was well-known in Trinity County, Texas, where she grew up.

In fact, Eleanor has a valuable book actually written by Jesse James, Jr., at the turn of the century. Much of this book is devoted to describing the "murder" of his father, with the remainder stating emphatically time and again that Jesse never broke the law and that any crimes he might have committed were merely Civil war-related activities with Quantrill's band of raiders.

Anyway, Eleanor and I have wondered if this book was part of a cover-up, that Jesse's son knew that his father was alive in Texas and wrote this book to help perpetuate the tale of cowardly murder.

One last personal story: the "Granbury Jesse James" had a grandson named Jesse Lee James, known as the "Hawk," who approached me several years ago with a complicated tale about a valuable treasure hidden by his outlaw grandfather. Hawk told me that the old man in Granbury had precisely located it for him. All he needed were Garrett metal detectors to help him recover it and funds to finance the search. I refused Hawk as I've similarly refused countless others over the years. My business is making metal detectors. Now, my hobby may be treasure hunting, but I look doubtfully at financing the adventures of others.

I understand that Hawk eventually found his "grubstake" partners, but the story ended as you probably suspected. They never recovered any hidden wealth. They squabbled over distribution rights and Hawk refused to lead the group to the treasure if indeed it ever existed.

Enough about me, let's talk about the legend of Jesse James and consider his numerous caches.

There's scarcely a state in the Central United States from Canada to Mexico that doesn't lay claim to having a cache of loot hidden by the James gang. I have been told "reliably" (and, usually, confidentially) about the exact location of Jesse James caches in eleven states with a total value of almost $3.5 million. He's even supposed to have robbed a mule train of silver or gold in Mexico. The 18 burro loads of Mexican gold (or silver) are supposedly buried in the mountains of Southwest Oklahoma. Perhaps this was the cache, among many others, that Hawk wanted me to help him recover!

Now, these so-called "facts" about Jesse James' caches are simply absurd. Making them even more ridiculous are their locations in places where he was never known to have visited. When Jesse, his brother Frank, the Younger brothers and those desperados who accompanied them were running wild through Missouri, Kansas, Nebraska, Iowa, and adjoining states, crimes literally throughout the nation were being attributed to them. There can be no question that Jesse and his gang were falsely blamed for countless misdeeds. On numerous occasions they were accused of robberies committed at the same time, yet hundreds of miles apart. When these facts were brought to the attention of authorities, the answer sometimes given was, "The

180

gang split up; Jesse planned all the crimes." Of notoriety, he had plenty!

A few years ago an enterprising researcher sought to list all the crimes that could be directly attributed to the James gang and to be liberal in estimating the amount of loot taken in each job. He came up with a total of 27 or 28 robberies that definitely were committed by Jesse and his gang. And, the total amount they took over a 16 year period amounted to less than $100,000.

Now, despite the claims of his son in Eleanor's book, there can be no doubt that Jesse and Frank James, the Younger brothers and accompanying renegades committed more than a couple dozen bank and train robberies. Yet the gang always had several members so there could never have been much money left after the loot from the various jobs was divided. Also, since the robberies were spaced out over 16 years, there's a good chance that all the money was spent on normal living expenses. There wouldn't have been much left to bury! It's been said by those who knew him that Jesse never had more than $6,000 in his possession at one time. Plus, it's a known fact that he was broke when Bob Ford "murdered" him; his widow had to auction off the household effects after the funeral. Remember, too, that Frank and Jesse spent two years farming in Tennessee. That would have been a mighty frustrating way to make a living for two fellows who knew the exact location of thousands of dollars in hidden caches.

What about all the "treasure maps" and stories that tell of finding treasure hidden by Jesse James? What about the literally thousands of man-hours that have been spent by treasure hunters in searching for Jesse's caches?

Well, Jesse became a legend during his lifetime, and that legend grew steadily after his death, whenever it occurred. When people let themselves get involved with legends, they ignore facts. I always remember the scene from that great Jimmy Stewart/John Wayne movie *The Man Who Shot Liberty Valance* when a Western newspaper editor counsels his reporter to "print the legend," even in the face of conflicting facts. Thus, an individual treasure hunter simply believes what he or she wants to believe. That's part of the mystery and beauty of our hobby!

I'm certainly not going to argue with a person who knows the exact location or even claims to have discovered a cache put down by Jesse James. Maybe he, Frank, Cole or another member of the gang actually hid some of their stolen money. Perhaps the James brothers really did bury millions of dollars worth of Mexican gold in a cave in the Oklahoma mountains. Far be it from me to try to disillusion any dedicated treasure hunter. If you have purchased a genuine treasure map or learned the secret of a Jesse James cache, I can only wish you the best of luck in finding it!

Chapter Twenty
Burying Caches

So, you want to bury your own treasure cache? You want it hidden so that nobody else can find it? Well, it's possible that you can achieve success because there are several I'm looking for that have virtually exhausted my resources. What is it that you want to hide? Most treasures that are waiting to be buried, as well as thousands that are already hidden away, represent fortunes, stolen loot and even ammunition caches.

How many· buried treasure caches are there in the world today, just waiting to be found? My guess is about ten million, and the actual number could be several times that large. I believe there's enough lost treasure lost in the world right now to easily pay off the national debt of the United States...with plenty of wealth left over to handle the debts of many smaller countries as well.

There are all kinds of little "gadgets" that you can buy in which to hide valuables in your own home. Perhaps you've seen these containers advertised in various magazines. If not, you can find them at most large hardware stores or various "spy shops" that have popped up lately. The hiding devices all have their purposes.

Is it possible to put down your treasure so it can't be found with a metal detector? A strange question for me to ask, you might think. Well, I could write a book on this subject and actually I have. My recent *Ghost Town Treasures* book is filled with tips on how to locate buried wealth and hide your own treasure. Believe me when I tell you, that book will certainly give you a lot to think about.

Let's consider the PVC-pipe treasure container. First, you must acquire a piece of PVC (plastic) pipe six to eight feet long and at least five inches in diameter. Next, dig a hole large enough to hold your pipe and about six inches deeper than its length. Then insert the pipe into the hole, with the top of the pipe six inches or more below ground surface. Put your treasure in a large plastic jar that will fit into the pipe, and seal the jar with a plastic lid.

Okay, how can you rig up your cache, which you will lower to the bottom of the pipe, so that you can recover it quickly and easily? I recommend that you attach to the top of the jar a length of non-metallic cord that will resist rot. The cord must be long enough to reach the bottom of the pipe. Next, secure the other end of the cord to the underside of a flat piece of material large enough to cover the upper end of the pipe. It should be made of plexiglass or a similar plastic or non-metallic substance. It will have to be strong enough to support the weight of a person, should someone ever stand on it. By lifting this "cover plate," you can easily remove your treasure in its container from the pipe.

Finally, place the "cover plate" on the top of the pipe and fill in the hole above it to ground level with dirt or matching soil. Your treasure is now safe and far enough beneath ground surface that a metal detector can't locate it. This method, of course, needs refinements, which you'll have to work out for yourself. If the water table is less than six feet deep, for example, you will have to seal off the bottom of your pipe to keep water from rising into it. Of course, rising water could force the pipe to rise up out of the hole. And, of course, you will have to mark the pipe's location in your memory very carefully or develop a memory code.

Even then, however, depending on ground mineralization and how the ground balance controls of a metal detector have been adjusted, your hole could possibly be located by a treasure hunter with "dumb luck." This is a "one-in-a-million" situation that is definitely not likely to happen but conceivably could.

What's another good way to cache a treasure? How about simply burying it in a junkyard! Even the most patient, dedicated and hard working treasure hunter will eventually walk away from a known treasure site if he has to dig tons and tons of junk metal to recover it.

Here's an example of what I mean. In the following chapter I list a few major treasure caches for which I've searched. One of these is a that of an East Texas doctor. We have a good idea about where he buried his treasure. A location where he kept many dogs to protect it. He fed them countless cases of canned dog food, and it seems that he must have thrown every last empty can into his one-acre back yard where the treasure was buried. How does a metal detector hobbyist stand a chance of locating a treasure when he has to compete with several thousand buried tin cans?

Good luck. I hope you are successful both in hiding your own treasure securely and in recovering one that has been lost and abandoned by another person, especially a desperado.

I really do. But, remember, I don't condone any man stealing another man's treasure.

Chapter Twenty-One
Caches I've Hunted

Writing this book has been a joy because it brought back many wonderful memories of my own adventures over the course of my lifetime. Over the past 50 years I've been involved in numerous searches, so many in fact, that I could never recall them all. I first thought that I would list, in this book, the most important, but where do I begin? And where do I draw the line?

So, I've only listed a few, as they came to mind, that I consider significant. There are many reasons for these treasure hunts to be memorable for me. Some of them were real heart-pounding "Indiana Jones-type" adventures. Others merely exhausted my physical energy as well as patience.

I was heavily involved in many of these searches. In others I worked only on the fringes. Some were solo activities; others, I was a part of a team. Some of the treasures were recovered; a great many were not. In a few heartbreaking instances we found only an empty hole to reward long and strenuous efforts.

Several of the searches are still in progress, at least a dozen. I am confident I know exactly where some of these treasures are located, but I cannot get to them for one reason or another. For example, foreign governments and the routes of narcotic traffic keep me away. The finder of some of these caches will be fabulously wealthy, while locating any one of them should make any hobbyist proud and quite jubilant.

For strictly personal and obviously, non-monetary, reasons I'd like to find the treasure cache of buttons I hid when I was a child. But I suspect it will stay buried because too many features of its location have changed. Also, the treasure is not metallic; I can't hope to locate it with my metal detector, no matter how hard I try.

Ask me about some of the treasures I *have* found, and I'll be glad to tell you about the flower-garden-treasure-pot of coins and the twin jackpot silver caches I recovered in Canada. I might even reminisce about a few more.

But, here's my list of treasure hunts for caches that I recall as being memorable:

The treasures of Nino Cochise; I'm a loner on this one. Why don't you read his book? Here's the title: *The First Hundred Years*, by Nino Cochise.

The stolen gold bars in Mexico; I almost froze to death one night while in this location.

Castle treasures in Spain, Germany and Scotland; great searching.

The 500-year-old inn treasure; I got a signal, and I'm still hoping to convince the tavern owner to rip into the location.

The dead man's cache east of the Hwy. 94 bridge, about eight or so miles west of my hometown of Lufkin, Texas; I've searched for it several times near the sagging fence which the old man crossed from the road to get to his silver-dollar treasure. Just west of the main bridge is a series of "beer joints". The fellow used to crawl through the north fence on the east side of the bridge, get some silver dollars, and then walk across the bridge toward the "beer joints".

The pig-sty treasure my father told me about, located about one mile west of downtown Lufkin on old Hwy. 94, northwest of a small bridge; my father said this one is a reality. I've dug a lot of junk iron there. The owner of the cache has been dead for 40 years.

Countless farmers' banks and farmers' wives butter-and-egg-money garden treasures; lots of opportunities here. The folks are all dead now but their treasures await the finder.

A cache of depression-era (1930s) coins under the Saturn Road bridge north of the Saturn Road Church of Christ and the Orchard Hills Baptist Church, halfway between Miller and Kingsley Roads in Garland, Texas; CCC workers hid them in concrete for good luck.

Several Jesse James treasures: in Oklahoma; near the Garrett factory; near the tri-state point of Texas, Louisiana and Arkansas; this last one is a spooky location to which I will never return. It's "guarded" by a bearded apparition who carries a long rifle and is a descendant of the Hatfield and McCoys.

U.S. buffalo soldier treasure near Ft. Davis, Texas; it even shoots!

$20 million (current gold coin value) stagecoach robbery stolen in the State of Washington; It's a matter of record; Roy Lagal and I know the rest of the story which is told on a Garrett video *Tracking Outlaw Treasure*.

Buried treasure near Ft. Pierce, Florida; 300-year-old Spanish coins come out of the sand looking freshly minted.

Fagley Lake deerskin treasure near Apple Springs, Texas; my wife's father said it's there.

Ghost town treasures near Terlingua, Texas; a bucket of silver dollars buried by twin brothers.

The Caribbean Island of Guadalupe treasure.

The cache of 1920s slot machine tokens I found. My father told me how they got there.

Buried treasure near Monahans, Texas; from Karl von Mueller's LUE map. I believe I've been close to it.

Chest of gold coins east of Boones Ferry on the Neches River; this one Judge Parker, of Woodville, Texas, helped me with.

Indian gold caches.

A long-deceased woman's treasure near Spalding, Idaho; $20 gold pieces.

McClain treasure on Hwy. 103, about five miles west of Lufkin; my friend's grandfather buried his money. It's probably still there.

Stolen safe hidden at a lake about four miles north of downtown Lufkin.

19th-century safe in a river near Reno, Nevada; George Mroczkowski and I worked on this one. It will take an aerial cable to locate it. I don't know how to walk on water, yet!

John Murrell treasures in Louisiana. I'm convinced some still exists for the finding and taking.

Helicopter search for stolen Ft. Worth bank loot.

Doctor's treasure located near Avenger, Texas; I had a waybill in my hands.

Houston doctor's *big* treasure. Gold and silver coins.

The silver tree in Mexico

At least three Pancho Villas treasures: in his home, in the nearby Batopillas, Mexico, ghost town and near San Antonio, Texas.

Aztec ceremonial pool treasure; near the Batopillas River near the cave of the vampire bats.

The twin jackpot silver caches I found in Cobalt, Ontario, Canada. Roy Lagal is my witness.

The cache of 400 B.C. coins I found in a plowed field in Greece.

Lost padre church wealth.

Ground fire coin caches in Mexico.

Adobe ghost treasure in Mexico; the little girl felt the ghost slap her bottom and he (it?) hacked down a door.

Seven tons of gold and silver south of Monahans in West Texas; one gold bar already found.

San Rafael and Robbers Roost, Utah, outlaw treasures. And, I wonder about Cache County, Utah.

Idaho kitchen sink treasure; we looked and looked and looked. Found a "tiny" amount of "Indian" money.

Idaho "put-down-for-keeps" treasures. Some of it was won in a poker game.

Idaho cave treasure; maybe in the storeroom of a national museum.

Twelve nickel slot machines; pitched into a lake by my father and grandfather in 1920. And they are filled with nickels!
190

Cache of World War II tokens near Cripple Creek, Colorado. Along a fence row.

One-cent coins from a bank robbery. My father told me about it.

Deathbed confession; this one's in the family. My wife knows all about it, except the exact spot where its buried

Grandfather's gold coins; also in the family. He buried it under the bee hive?

Good luck to you in your cache hunting endeavors. I know you can find them, because I have seen and heard about a lot more than I have searched for and found!

Oh, wait! There's one more:

I've already told you about my button cache I buried about the time I started grade school. I have yet to find it, but someday hope for the chance. Later, I saw the movie "Treasure Island" with Wallace Berry. Boy, I became excited when I saw that pirate's treasure that was hidden in a cave. I knew right then that someday I wanted to find my own buried treasure cache.

A few years later I had my chance. A local radio station announced that they had concealed a treasure that anyone could own if they could find it–a check for $100 (a lot of money for that time) hidden somewhere in Lufkin, Texas. My friend Edgar McClain and I jumped into the search. Each day we listened for the announcer to reveal another clue to the whereabouts of the treasure. I remember a few of them: "Back and forth, back and forth, will it never stop?" "Not six high, not low, just in between." "Between here and there, a half mile, for sure." "One turn, two turns, three turns and you got it."

Edgar and I spent considerable time studying the clues. We walked and drove all over town searching and searching and searching. Alas, the day came when the winner was announced. It wasn't Edgar and I.

But, after all these years I remember the check was located a half mile from the radio station at a point adjacent to a street where traffic continually drove "back and forth, back and forth, will it never stop?"

The check was concealed inside a three-inch diameter pipe that was protruding vertically about three feet above the sidewalk. A pipe end cap was turned "one turn, two turns, three turns" and who reached in and pulled out the check, I don't remember.

I do remember, however, that after many, many years I still have strong, never-ending desire to be involved in the search for treasure. When the treasure bug bites, you're bitten!

Of course, I agree wholeheartedly with what Gar Starrett always said in my novels: *If nobody gets hurt or spends money he or she can't afford, every treasure hunt is a genuine pleasure.*

So, maybe when you're out someday enjoying your pleasure as you use a metal detector and a genuine treasure map to search for a cache that you know was hidden by Jesse James or some other famous outlaw.

I'll see you in the field...
and God bless, again!

RAM Books
Order Form

Please send me the following RAM Books:

- ☐ ⁺New Successful Coin Hunting...............................$9.95
- ☐ Treasure Hunting for Fun and Profit........................$9.95
- ☐ Ghost Town Treasures...$9.95
- ☐ Find Gold with a Metal Detector..............................$9.95
- ☐ Buried Treasures You Can Find.............................$14.95
- ☐ Gold of the Americas..$9.95
- ☐ Treasure Recovery From Sand and Sea................$7.95
- ☐ New Modern Metal Detectors...............................$12.95
- ☐ Gold Panning is Easy..$9.95
- ☐ Introduction to Metal Detecting*............................$1.00
- ☐ Competitive Treasure Hunting...............................$9.95
- ☐ Find an Ounce of Gold a Day.................................$3.00

Add $1.00 for each book. (Maximum of $3.00) for shipping and handling.
*No $1.00 shipping charge when ordered with another book.

Send order form and payment to:
Garrett Metal Detectors
RAM Publishing
1881 West State Street
Garland, Texas 75042

Total book purchase amount..............$ _____

8.25% tax (Texas residents only).......$ _____

Shipping and Handling....................$ _____

Total...$ _____

Please check one
☐   Enclosed is my check or money order
☐   I prefer to pay using my credit card (check one)
        ☐ American Express   ☐ MasterCard
        ☐ Visa   ☐ Discover

Please fill out the following information:

_____
Credit Card Number (include expiration date)

_____
Name

_____
Shipping Address

_____
City, State, Zip Code

_____
Signature (all credit card orders must be signed)